Viacheslav Emelyanov

Dynamical approach to the cosmological constant problem

Viacheslav Emelyanov

Dynamical approach to the cosmological constant problem

Südwestdeutscher Verlag für Hochschulschriften

Impressum / Imprint
Bibliografische Information der Deutschen Nationalbibliothek: Die Deutsche Nationalbibliothek verzeichnet diese Publikation in der Deutschen Nationalbibliografie; detaillierte bibliografische Daten sind im Internet über http://dnb.d-nb.de abrufbar.

Bibliographic information published by the Deutsche Nationalbibliothek: The Deutsche Nationalbibliothek lists this publication in the Deutsche Nationalbibliografie; detailed bibliographic data are available in the Internet at http://dnb.d-nb.de.

Coverbild / Cover image: www.ingimage.com

Verlag / Publisher:
Südwestdeutscher Verlag für Hochschulschriften
ist ein Imprint der / is a trademark of
OmniScriptum GmbH & Co. KG
Heinrich-Böcking-Str. 6-8, 66121 Saarbrücken, Deutschland / Germany
Email: info@svh-verlag.de

Herstellung: siehe letzte Seite /
Printed at: see last page
ISBN: 978-3-8381-3573-1

Zugl. / Approved by: Karlsruhe, Karlsruhe Institute of Technology, Diss., 2012

Contents

Conventions and Abbreviations

We work throughout in units implying $\hbar = c = 1$, where $\hbar$ is the reduced Planck constant and c is the speed of light in vacuum.

In what follows, we consider that the universe is a four-dimensional, globally hyperbolic manifold M with a metric g. We specify every point on M as $x \equiv x^\mu = (x^0, x^1, x^2, x^3)$, where x^0 is time and x^i, where $i = 1, 2, 3$ are spatial coordinates. Greek indices always run over 0, 1, 2, 3, while roman ones denote only spatial components.

We designate the metric components by $g_{\mu\nu}$, i.e. $g = g_{\mu\nu}\, dx^\mu \otimes dx^\nu$, with the signature $(+, -, -, -)$. At certain points in this work, we will use two special configurations of $g_{\mu\nu}$, namely $\delta_{\mu\nu}$ and $\eta_{\mu\nu}$. The former represents the metric of the four-dimensional Euclidean space E^4, and the latter is the metric of Minkowski spacetime $M^4_{1,3}$.

The affine connection ∇ is assumed to be the Levi-Civita connection. Thus the connection coefficients ${\Gamma^\lambda}_{\mu\nu}$ are given by the Christoffel symbols:

$$ {\Gamma^\lambda}_{\mu\nu} = \frac{1}{2} g^{\lambda\rho} \left(\partial_\mu g_{\nu\rho} + \partial_\nu g_{\mu\rho} - \partial_\rho g_{\mu\nu} \right), $$

where ∂_μ is the partial derivative with respect to x^μ, i.e. $\partial / \partial x^\mu$. We will frequently use shorthand notations $(...)_{;\,\mu}$ and $(...)_{,\mu}$ for denoting the covariant and partial derivatives, respectively.

The components of the Riemannian curvature tensor ${R^\mu}_{\nu\lambda\rho}$ are given by

$$ {R^\mu}_{\nu\lambda\rho} \equiv \partial_\lambda {\Gamma^\mu}_{\nu\rho} - \partial_\rho {\Gamma^\mu}_{\nu\lambda} + {\Gamma^\mu}_{\sigma\lambda} {\Gamma^\sigma}_{\nu\rho} - {\Gamma^\mu}_{\sigma\rho} {\Gamma^\sigma}_{\nu\lambda} \, . $$

The Ricci tensor and scalar are defined as $R_{\mu\nu} \equiv {R^\lambda}_{\mu\lambda\nu}$ and $R \equiv g^{\mu\nu} R_{\mu\nu}$, respectively.

Chapter 1

Introduction

Recent observations of type Ia supernovae have shown that the expansion of the universe is accelerating [1]. This phase has started at redshift $z \sim 1$, that corresponds to the epoch of galaxy formation. These facts were later confirmed by other experiments [2, 3]. This phenomenon goes under the name "dark energy". The outstanding question is to discover the physics of it.

The up-to-date experimental data show that its energy density $\rho_{\rm de}$ is approximately equal to 72.8% of the critical mass density:

$$\rho_{\rm de} \approx 2.9 \times 10^{-47}\ ({\rm GeV})^4 ,$$

the pressure $P_{\rm de}$ of it is negative, with the absolute value roughly equaling its energy density, i.e. $P_{\rm de} \approx -\rho_{\rm de}$.

Many proposals have been suggested in an attempt to uncover its nature, among which is the cosmological constant associated with the lowest-energy state of quantum fields. The fact is that quantum theory implies that the normal or ground state of a certain field is endowed with the zero-point or vacuum energy density $\rho_{\rm V}$, which vanishes in the classical limit $\hbar \to 0$. Vacuum also possesses the pressure $P_{\rm V}$, that is related to its energy density as $P_{\rm V} = -\rho_{\rm V}$.

Naive theoretical estimates give an unacceptably large value of the zero-point energy density $\rho_{\rm V}$, that strongly contradicts the observations. In addition, the nonzero Higgs condensate in the standard electroweak theory and the quark and gluon condensates in quantum chromodynamics make enormous contributions to the total vacuum energy as well. This is the essence of the cosmological constant problem (CCP).

In addition to this puzzle, there are two related cosmological problems. Specifically, it is asked why $\rho_{\rm de}$ is not precisely zero, but of the order of the matter energy density $\rho_{\rm m}$ of the universe, taking into account that $\rho_{\rm de}$ and $\rho_{\rm m}$ depend differently on cosmic time. The latter is the so-called cosmic coincidence problem.

In the present research, we address the question of how to dynamically get rid of the large value of the total vacuum energy produced by quantum fields.

This thesis is arranged as follows: in Chapter 2, we will briefly describe our theoretical framework that will be used throughout this work. Then, in Chapter 3, we will talk over the first cosmological constant problem in detail. In Chapter 4, particular approaches to the problem are briefly discussed. These are fine-tuning, dynamical adjustment and q-theory [4]. Motivated by q-theory and its special realization [5, 6], we will treat in Chapter 5 vector-tensor model giving a dynamical cancellation of the total vacuum energy appearing in the Einstein equations. We will also discuss a serious obstacle inherent in it. In Chapter 6, we will considerably modify this model in order to overcome that flaw. And, finally, in Chapters 7 and 8, we will discuss our results and conclude.

Chapter 2

Theoretical framework

According to contemporary physics, there exist four fundamental interactions in nature. These are the electromagnetic, weak, strong and gravitational forces.

The first three interactions are well-described by gauge vector bosons according to the standard model of particle physics based on quantum field theory with the local symmetry group $U(1)_{\mathrm{Y}} \times SU(2)_{\mathrm{L}} \times SU(3)_{\mathrm{C}}$ [7, 8].

General relativity (GR) is a classical theory of gravity. This theory is invariant under Diff(M) group and based on geometrical and dynamical (equivalence) principles [9]. The geometrical idea implies spacetime is a torsion-free manifold that locally looks as $M^4_{1,3}$. Thus the metric g contains all information about the gravity. The equivalence principle means that inertial mass of a body coincides with the gravitational one.

The weak and strong interactions are short-range, while the electromagnetism and gravity are long-range. However, matter is electrically neutral on average, so that gravity governs the evolution of the universe on large scales.

2.1 Einstein-Hilbert action

The dynamics of the metric in general relativity is determined by

$$S_{\mathrm{EH}}[g] \;=\; -\frac{1}{16\pi G}\int d^4x\sqrt{-g}\,R\,, \tag{2.1}$$

where the integration is performed over the manifold M with $\partial M = 0$. This action was proposed by Hilbert in 1915 and is known as the Einstein-Hilbert action functional, where G is the gravitational constant [9].

The Einstein field equations are derived by setting the functional derivative of (2.1) over $g_{\mu\nu}$ to zero. Thus one has

$$G_{\mu\nu} \;\equiv\; R_{\mu\nu} - \frac{1}{2}R\,g_{\mu\nu} \;=\; 8\pi G\,T^{\mathrm{m}}_{\mu\nu}\,, \tag{2.2}$$

where we have also added a matter action $S^{\rm m}[g,\psi]$ to (2.1) with an energy-momentum tensor (EMT) defined in the usual manner:

$$T^{\rm m}_{\mu\nu}(x) \;\equiv\; \frac{2}{\sqrt{-g}}\frac{\delta S^{\rm m}[g,\psi]}{\delta g^{\mu\nu}(x)}\,, \tag{2.3}$$

and $G_{\mu\nu}$ in (2.2) is the Einstein tensor.

2.2 Robertson-Walker metric

Observational data show that the universe is homogeneous and isotropic on scales larger than 100 Mpc. The metric tensor g of such the universe is

$$dt\otimes dt - a^2(t)\left(\frac{dr\otimes dr}{1-kr^2} + r^2\Big(d\theta\otimes d\theta + \sin^2\theta\, d\varphi\otimes d\varphi\Big)\right) \tag{2.4}$$

and known as the Robertson-Walker metric, where $a(t)$ is a scale factor and the curvature constant $k \in \{-1,0,+1\}$. These values of k correspond to open, flat and closed universes, respectively. Observations indicate that our universe is roughly flat, therefore we set $k=0$ in what follows.

Having used (2.4) with $k=0$, one finds the nonzero components of $G_{\mu\nu}$:

$$G_{00} \;=\; 3H^2 \quad\text{and}\quad G_{ij} \;=\; (2\dot{H}+3H^2)g_{ij}\,, \tag{2.5}$$

where dot stands for a differentiation over t and $H \equiv \dot{a}/a$ is the Hubble parameter. Note, the symmetries of M encoded in (2.4) imply $T^{\rm m}_{0i}=0$ and $T^{\rm m}_{ij}\propto g_{ij}$ as these directly follow from (2.2) and (2.5).

2.3 Matter composition of the universe

Cosmic microwave background radiation The cosmic microwave background (CMB) comprises primordial photons that decoupled from matter after the recombination epoch $z_{\rm rec}\approx 1100$. The CMB has the Planck spectrum with a temperature $T\approx 2.725$ K and anisotropies at the level of 10^{-5} [10].

The electromagnetic field is mathematically described by a one-form $A = A_\mu dx^\mu$, the dynamics of which is governed by $\mathcal{L}_{\rm EM} = -\frac{1}{4}F_{\mu\nu}F^{\mu\nu}$, where $F_{\mu\nu} = 2\nabla_{[\mu}A_{\nu]}$ [11]. Hence its energy-momentum tensor is in components

$$T^{\rm rad}_{00} \;=\; \frac{1}{2a^2}\left(|\mathbf{E}|^2+\frac{|\mathbf{B}|^2}{a^2}\right), \tag{2.6}$$

$$T^{\rm rad}_{0i} \;=\; \frac{1}{a^2}(\mathbf{E}\times\mathbf{B})_i\,, \tag{2.7}$$

$$T^{\rm rad}_{ij} \;=\; -\left(E_iE_j+\frac{B_iB_j}{a^2}+\frac{1}{2a^2}\left(|\mathbf{E}|^2+\frac{|\mathbf{B}|^2}{a^2}\right)g_{ij}\right), \tag{2.8}$$

where by definition $E_i \equiv F_{0i}$ and $B_i \equiv \varepsilon_{ijk}F_{jk}$ are electric and magnetic fields, ε_{ijk} is the three-dimensional Levi-Civita symbol and $|\mathbf{E}|^2 \equiv \delta^{ij}E_iE_j$, $|\mathbf{B}|^2 \equiv \delta^{ij}B_iB_j$.

Taking into account the isotropy of the CMB, we average $T^{\rm rad}_{\mu\nu}$ over the solid angle. This implies $\langle(\mathbf{E}\times\mathbf{B})_i\rangle = 0$, $\langle E_i E_j\rangle = \frac{1}{3}|\mathbf{E}|^2\delta_{ij}$ and $\langle B_i B_j\rangle = \frac{1}{3}|\mathbf{B}|^2\delta_{ij}$, where the angle brackets denote the average over the angles. We have

$$\langle T^{\rm rad}_{00}\rangle = \frac{1}{2a^2}\left(|\mathbf{E}|^2+\frac{|\mathbf{B}|^2}{a^2}\right) \quad \text{and} \quad \langle T^{\rm rad}_{ij}\rangle = -\frac{1}{6a^2}\left(|\mathbf{E}|^2+\frac{|\mathbf{B}|^2}{a^2}\right) g_{ij}\,. \tag{2.9}$$

Since $\langle T^{\rm rad}_{00}\rangle = \rho_{\rm rad}$ and $\langle T^{\rm rad}_{ij}\rangle = -P_{\rm rad}g_{ij}$, we find $P_{\rm rad} = \rho_{\rm rad}/3$.

Baryonic and dark matter Baryonic matter is a matter composed of normal atomic matter.

Dark matter is an unknown component of the universe, that was originally introduced in order to explain a large value of the mass-to-light ratio in galaxies and clusters of galaxies [12, 13]. Observations indicate that dark matter energy density is roughly 22.7% of the critical mass density and has negligible pressure. There are currently several dark matter candidates, such those weakly interacting massive particles, sterile neutrinos, axions and others [14].(1)

On sufficiently large scales, matter can be roughly regarded as a perfect fluid [9, 10] characterized by its energy density ρ, pressure P and four-velocity u_μ with

$$T_{\mu\nu} = (\rho+P)u_\mu u_\nu - P g_{\mu\nu}\,. \tag{2.10}$$

For a non-relativistic, dust-like perfect fluid ($u_0 \approx 1$, $|\mathbf{u}| \ll 1$, $P_{\rm dust} \ll \rho_{\rm dust}$), one has

$$T^{\rm dust}_{00} \approx \rho_{\rm dust}\,, \quad T^{\rm dust}_{ij} \approx -P_{\rm dust}g_{ij}\,. \tag{2.11}$$

In what follows, the baryonic and dark matters are considered as such kind of the fluid with the zero pressure.

Dark energy As mentioned above, observations indicate $P_{\rm de} \approx -\rho_{\rm de}$. From now on we also assume that this is the exact equality. Substituting this in (2.10), we obtain

$$T^{\rm de}_{00} = \rho_{\rm de}\,, \quad T^{\rm de}_{ij} = \rho_{\rm de}g_{ij}\,. \tag{2.12}$$

2.4 Cosmological evolution of the universe

Taking (2.2) with (2.5) and (2.9), (2.11) as well as (2.12), we obtain one of the Friedmann equations describing the evolution of the universe:

$$\dot{a}^2 = \frac{8\pi}{3}G\,a^2\Big(\rho_{\rm de}+\rho_{\rm rad}+\rho_{\rm dust}\Big)\,. \tag{2.13}$$

(1)As an alternative, Milgrom suggested to modify the Newtonian dynamics in order to resolve this missing mass problem. See the original paper [15] for details.

It follows from the Bianchi identities that $\nabla^\mu G_{\mu\nu} = 0$. On the other hand, $\nabla^\mu T^{\rm m}_{\mu\nu} = 0$ holds along the matter field equations. These are

$$\dot{\rho}_{\rm rad} = -4H\rho_{\rm rad}\,, \quad \dot{\rho}_{\rm dust} = -3H\rho_{\rm dust}\,, \quad \dot{\rho}_{\rm de} = 0\,. \tag{2.14}$$

Since $H = \dot{a}/a$, we find

$$\rho_{\rm rad} = \rho_{\rm rad0}\left(\frac{a_0}{a}\right)^4, \quad \rho_{\rm dust} = \rho_{\rm dust0}\left(\frac{a_0}{a}\right)^3, \quad \rho_{\rm de} = \rho_{\rm de0}\,, \tag{2.15}$$

where the subscript 0 near ρ and a denotes their values at the present day $t_0 \sim H_0^{-1}$.

After a possible epoch of inflation [16, 17], when the universe was small and hot, matter was ultra-relativistic. This is the so-called radiation-dominated phase. Setting $\rho_{\rm de}$ and $\rho_{\rm dust}$ to zero, the Friedmann equation gives $a(t) \propto t^{1/2}$ ($\dot{a} > 0$, $\ddot{a} < 0$).

Then the universe cooled down due to its expansion and massive matter became non-relativistic with an energy density larger than that of radiation – a dust-dominated phase which took place after roughly 10^{11} sec after big bang. Taking into account only $\rho_{\rm dust}$ and neglecting others in (2.13), one deduces $a(t) \propto t^{2/3}$ ($\dot{a} > 0$, $\ddot{a} < 0$).

At redshift $z \sim 1$ corresponding to the epoch of galaxy formation, dark energy has started to dominate at large scales. Neglecting $\rho_{\rm rad}$ and $\rho_{\rm dust}$ in comparison with $\rho_{\rm de}$, one obtains from (2.13) that $a(t) \propto \exp(Ht)$ ($\dot{a} > 0$, $\ddot{a} > 0$), where $H = (8\pi G\rho_{\rm de}/3)^{1/2}$.

2.5 Newton's law of gravity

General relativity reduces to Newtonian gravity in the limit of a weak gravitational field and small velocities corresponding to $c \to \infty$.

The weak gravitational field means the metric tensor is close to Minkowski one:

$$g_{\mu\nu}(x) = \eta_{\mu\nu} + h_{\mu\nu}(x)\,, \quad \text{where} \quad |h_{\mu\nu}(x)| \ll 1\,. \tag{2.16}$$

This can always be achieved by using the Riemannian normal coordinates $\bar{x}^\mu$, in which $g_{\mu\nu}(\bar{x})$ roughly becomes $\eta_{\mu\nu} + \frac{1}{3}R_{\mu\lambda\rho\nu}\bar{x}^\lambda\bar{x}^\rho$ [9].

Einstein's equations linearized with respect to the metric perturbations $h_{\mu\nu}(x)$ are

$$\partial^2 h_{\mu\nu} - \partial_\mu\partial^\lambda h_{\nu\lambda} - \partial_\nu\partial^\lambda h_{\mu\lambda} + \eta^{\lambda\rho}\partial_\mu\partial_\nu h_{\lambda\rho} + 2\kappa^2\left(\delta T^{\rm m}_{\mu\nu} - \frac{1}{2}\eta_{\mu\nu}\delta T^{\rm m}\right) = 0\,, \tag{2.17}$$

where by definition $\delta T^{\rm m} \equiv \eta^{\lambda\rho}\delta T^{\rm m}_{\lambda\rho}$ and $\kappa^2 \equiv 8\pi G$ is the Einstein gravitational constant introduced here for the sake of convenience.

Since GR is Diff(M) invariant, (2.17) must be invariant under a coordinate transformation $x^\mu \to \tilde{x}^\mu = x^\mu - \xi^\mu$, where ξ^μ is an infinitesimally small function of a given spacetime point. It is straightforward to show that (2.17) is invariant under the replacement $h_{\mu\nu} \to h_{\mu\nu} + \mathcal{L}_\xi\eta_{\mu\nu}$, where $\mathcal{L}_\xi$ is the Lie derivative along $\xi = \xi^\mu\partial_\mu$. This transformation enables to fix 4 functions.

In the harmonic gauge [9], (2.17) becomes

$$\partial^2 h_{\mu\nu} \;=\; -2\kappa^2 \left(\delta T^{\rm m}_{\mu\nu} - \frac{1}{2}\eta_{\mu\nu}\delta T^{\rm m} \right), \qquad \partial^2 \;\equiv\; \eta^{\mu\nu}\partial_\mu \partial_\nu \,. \tag{2.18}$$

A general solution of this equation with the omitted source term ($\partial^2 h_{\mu\nu} = 0$) describes a propagation of gravitational waves in empty space.

As noticed above, Newton's law of gravity corresponds to the case of the weak gravitational field with non-relativistic bodies. The energy-momentum tensor of such a body with a mass M at a given point $\mathbf{x}_0$ is $\delta T^{\rm m}_{\mu\nu} = M \delta^0_\mu \delta^0_\nu \delta(\mathbf{x} - \mathbf{x}_0)$, so that we immediately find from (2.18)

$$h_{\mu\nu}(\mathbf{r}) \;=\; -\frac{2GM}{|\mathbf{r}|}\,\delta_{\mu\nu}\,, \qquad \text{where} \quad \mathbf{r} \;\equiv\; \mathbf{x} - \mathbf{x}_0 \,. \tag{2.19}$$

Let us consider a freely-falling body m influenced by the gravitational field of M. According to general relativity, it moves along its geodesic $\nabla_u u = 0$, where $u = u^\mu \partial_\mu$ is its four-velocity. Taking into account $u^\mu \approx (1, \mathbf{u})$ and $|\mathbf{u}| \ll 1$, one has

$$m\ddot{\mathbf{r}} \;\approx\; -\frac{GmM}{|\mathbf{r}|^2} \cdot \frac{\mathbf{r}}{|\mathbf{r}|} \,. \tag{2.20}$$

This is Newton's famous law of gravity. Hence, G can be identified with Newton's gravitational constant $G_{\rm N} \approx 6.67 \times 10^{-11}$ m^3 kg^{-1}s^{-2} [18] in general relativity.

The Newton gravitational law is experimentally verified from 10^{-4} m [19] up to the size of solar system 10^{12} m.

Chapter 3

Cosmological constant problem

In 1917 Einstein introduced an extra constant term Λ into $\mathcal{L}_{\text{EH}}$, i.e.

$$S[g] = -\frac{1}{16\pi G_{\text{N}}} \int d^4x \sqrt{-g} \left(R + 2\Lambda \right), \tag{3.1}$$

which is known as the cosmological constant [20, 21]. His original goal was to have a static universe, however it turned out that the cosmological constant Λ does not provide a stable stationary solution. In addition, it was later discovered by Hubble that our universe is actually expanding [22].

The Einstein equations with the cosmological term read

$$R_{\mu\nu} - \frac{1}{2} R\, g_{\mu\nu} - \Lambda\, g_{\mu\nu} = 8\pi G_{\text{N}} T^{\text{m}}_{\mu\nu}, \tag{3.2}$$

where we have also added the matter field with the energy-momentum tensor $T^{\text{m}}_{\mu\nu}$.

One can associate an energy-momentum tensor $T^{\Lambda}_{\mu\nu}$ with the cosmological constant: $T^{\Lambda}_{\mu\nu} = \rho_\Lambda g_{\mu\nu}$, where by definition $\rho_\Lambda \equiv \Lambda / 8\pi G_{\text{N}}$ and, hence, $P_\Lambda = -\rho_\Lambda$. Thus, Λ can be regarded as one of the candidates for the explanation of the accelerated expansion of our universe.(1) Henceforth, we assume that this is the case, and the phrases dark energy and cosmological constant will be used interchangeably throughout our work.

We note that Λ can have any value from the point of view of general relativity.

3.1 Zero-point energy of quantum fields

In 1967 Zel'dovich pointed out that the energy density of a quantum field in its ground state could be related with the cosmological constant [26].

Let us consider $\phi : M^4_{1,3} \to \mathbb{R}$ with an action functional

$$S[\phi] = \int d^4x \left(\frac{1}{2} \eta^{\mu\nu} \partial_\mu \phi \partial_\nu \phi - \frac{1}{2} m^2 \phi^2 \right). \tag{3.3}$$

(1)There have also been proposed alternative ideas of how to explain the late accelerated expansion of the universe in the literature: quintessence [23], k-essence [24], $f(R)$ theories [25] and others.

where $m > 0$ is a constant standing for the mass of the field.

The dynamical variables are $\phi(x)$ and $\pi(x) \equiv \delta L/\delta\dot{\phi}(x) = \dot{\phi}(x)$ ($L[\phi] = \int d^3\mathbf{x}\,\mathcal{L}$ is the Lagrangian), which satisfy the well-known Hamilton equations. After quantization [8], they become operators defined on the Hilbert space. For instance, $\hat{\phi}(x)$ is given by

$$\hat{\phi}(x) = \int \frac{d^3\mathbf{k}}{(2\pi)^3}\frac{1}{\sqrt{2\omega_\mathbf{k}}}\left(\hat{a}_\mathbf{k}e^{-ikx} + \hat{a}^\dagger_\mathbf{k}e^{+ikx}\right), \tag{3.4}$$

where $k^\mu \equiv (\omega_\mathbf{k}, \mathbf{k})$ with $\omega_\mathbf{k} \equiv \sqrt{|\mathbf{k}|^2 + m^2}$ and

$$\left[\hat{a}_\mathbf{k}, \hat{a}^\dagger_{\mathbf{k}'}\right] = (2\pi)^3\delta(\mathbf{k}-\mathbf{k}'), \quad \left[\hat{a}_\mathbf{k}, \hat{a}_{\mathbf{k}'}\right] = \left[\hat{a}^\dagger_\mathbf{k}, \hat{a}^\dagger_{\mathbf{k}'}\right] = 0. \tag{3.5}$$

A state $|0\rangle$ is defined as the vacuum state, such that $\hat{a}_\mathbf{k}|0\rangle = 0\ \forall\,\mathbf{k}$. A state $\hat{a}^\dagger_\mathbf{k}|0\rangle$ represents a state with one excited $\mathbf{k}$-mode and so on. The vacuum state $|0\rangle$ is interpreted as a no-particle state, while $\sqrt{2\omega_\mathbf{k}}\hat{a}^\dagger_\mathbf{k}|0\rangle$ represents a particle with a four-momentum k^μ.

The vacuum expectation value of the Hamiltonian operator $\hat{H} = \int d^3\mathbf{x}\ \hat{T}_{00}$ is known as the zero-point energy E_0, i.e. $E_0 \equiv \langle 0|\hat{H}|0\rangle$. This energy is divergent, at the very least, because of the infinite volume of the three-dimensional space. If we consider a zero-point energy density, then this divergence is eliminated:

$$\rho_\text{V} \equiv \lim_{V\to\infty}\left(\frac{E_0}{V}\right) = \frac{1}{2}\int \frac{d^3\mathbf{k}}{(2\pi)^3}\,\omega_\mathbf{k}\,. \tag{3.6}$$

The vacuum expectation value of the momentum operator $-\langle 0|\int d^3\mathbf{x}\ \hat{T}_{0i}|0\rangle$ is precisely zero, but the vacuum expectation value of the pressure operator is given by(2)

$$P_\text{V} = \frac{1}{6}\int \frac{d^3\mathbf{k}}{(2\pi)^3}\frac{|\mathbf{k}|^2}{\omega_\mathbf{k}}\,. \tag{3.7}$$

Both integrals (3.6) and (3.7) are ultraviolet divergent. To prove $P_\text{V} = -\rho_\text{V}$, let us regularize P_V in the following way

$$P_\text{V}^\varepsilon = \frac{1}{3}\int\limits_0^{+\infty} \frac{k^4\exp(-\varepsilon k)}{\sqrt{k^2+m^2}}\,dk\,, \tag{3.8}$$

where $\varepsilon > 0$, so that $P_\text{V} = \lim\limits_{\varepsilon\to 0} P_\text{V}^\varepsilon$. Integrating P_V^ε by parts, and then considering the limit $\varepsilon \to 0$, one finds that ρ_V and P_V are similarly related as the corresponding quantities of the cosmological constant.(3)

In quantum field theories of particle physics, the zero-point energy does not play a significant role, since only energy differences have a physical meaning in scattering processes. Therefore, it is simply removed there by a normal ordering [8, 28]. It should be mentioned, however, that the elimination of the

(2)By this we mean a quantity $P_\text{V}\eta_{ij} = -\lim\limits_{V\to\infty}\left(V^{-1}\langle 0|\int d^3\mathbf{x}\ \hat{T}_{ij}\,|0\rangle\right)$.

(3)The regularization procedure presented here differs from that that Zel'dovich used, namely he suggested a Pauli-Villars regularization [27] of all divergences by introducing a spectrum of massive regulator fields [26].

zero-point energy by this prescription does not imply that vacuum fields has no physical consequences and, moreover, it is needed there for a formal mathematical consistency of the theory [29].

The QED effects such as the spontaneous emission [29], the Lamb shift [30, 31], the anomalous moment [32, 33] and the Casimir force [34, 35, 36, 37] are experimentally observed and, hence, serve as evidences for the reality of vacuum fluctuations. However, there are no laboratory evidences that the zero-point energy is real [38].(4)

3.2 Semi-classical Einstein equations

Quantum field theories are formulated in Minkowski spacetime $M^4_{1,3}$. Although the universe can always be locally regarded as $M^4_{1,3}$, this is not the case globally due to the curvature of the universe and its possible nontrivial topology. As a consequence, there is not, in general, a decomposition of the field ϕ into positive- and negative-frequency modes as in (3.4) and the concept of physical vacuum, as defined above, loses its unambiguous meaning as well as the concept of particle [40].

However, the energy-momentum tensor $T_{\mu\nu}$ of a quantum field appearing in the Einstein equations are defined locally and can be found by using the path integral method proposed by Feynman, while gravity is regarded as a classical field [40, 41].

Let us consider the free, real scalar field ϕ in a curved spacetime. Its action functional is given in (3.3) up to the replacement of d^4x and $\eta^{\mu\nu}$ by $d^4x\sqrt{-g}$ and $g^{\mu\nu}$, respectively. The vacuum expectation value of its energy-momentum tensor is

$$\langle \hat{T}_{\mu\nu} \rangle \equiv \frac{\int \mathcal{D}\phi \, T_{\mu\nu}(\phi) \exp(iS[\phi, g])}{\int \mathcal{D}\phi \exp(iS[\phi, g])} = \frac{2}{\sqrt{-g}} \frac{\delta \Gamma[g]}{\delta g^{\mu\nu}}, \tag{3.9}$$

where $\Gamma[g]$ is the effective action defined as $-i \langle 0_{\text{out}} | 0_{\text{in}} \rangle = -i \ln \int \mathcal{D}\phi \exp\left(iS[\phi, g]\right)$.(5) This integral is taken over ϕ satisfying certain boundary or periodicity conditions.

After variation an effective action of the whole system, i.e. the metric plus scalar fields, the semi-classical Einstein equations can be obtained:

$$R_{\mu\nu} - \frac{1}{2} R \, g_{\mu\nu} = \Lambda_0 \, g_{\mu\nu} + 8\pi G_0 \, \langle \hat{T}_{\mu\nu} \rangle , \tag{3.10}$$

which take into account the backreaction of quantum fluctuations of ϕ on the metric.

It is shown in Appendix A by following well-known methods, that $\langle \hat{T}_{\mu\nu} \rangle$ can be written in the weak-field limit as

$$\langle \hat{T}_{\mu\nu} \rangle = A(\tilde{\mu}, d) \, g_{\mu\nu} - 2B(\tilde{\mu}, d) \, G_{\mu\nu} + \mathrm{O}(g^2) , \tag{3.11}$$

where $A(\tilde{\mu}, d)$ and $B(\tilde{\mu}, d)$ are given in (9), d is the dimension of spacetime and $\tilde{\mu}$ is the t' Hooft scale.

(4)By saying zero-point or vacuum energy we mean vacuum bubbles which have no external legs (as in Figure 3.1 on page 17) in contrast to vacuum fluctuations (see [39]).

(5)We note that $\langle 0_{\text{out}} | 0_{\text{in}} \rangle$ is related to both $\langle 0_{\text{in}} | 0_{\text{in}} \rangle$ and $\langle 0_{\text{out}} | 0_{\text{out}} \rangle$ [40]. See also discussion of this issue in [41].

Substituting (3.11) in (3.10), one sees that Λ_0 and G_0 become Λ and G according to

$$\Lambda \equiv \Lambda_0 + 8\pi G_0 A(\tilde{\mu}, d) , \quad G \equiv \frac{G_0}{1 + 16\pi G_0 B(\tilde{\mu}, d)} , \tag{3.12}$$

and since, for instance, $A(\tilde{\mu}, d) \to \infty$ when d approaches 4, Λ_0 must also be divergent in the same manner as $A(\tilde{\mu}, d)$ in order for Λ appearing in the semi-classical Einstein equations (3.10) to be finite.

Applying dimensional regularization to the integral (3.6) (or (3.7)), one obtains $\rho_{\rm V} = A(\tilde{\mu}, d)$ (or $P_{\rm V} = -A(\tilde{\mu}, d)$), i.e. $\Lambda = \Lambda_0 + 8\pi G_0 \rho_{\rm V}$. Note, after renormalization, Λ and G correspond to the observed values of the cosmological constant and Newton's constant, respectively [40, 41].

The conclusion is that the zero-point energy of quantum fields gravitates in general relativity. Thus we cannot simply get rid of this as it is the case in quantum field theories in Minkwoski spacetime.

3.3 Cosmological constant problem

As mentioned in the Introduction, the problem comes from a comparison of the observed energy density of the cosmological constant with its theoretical estimates. These estimates contradict observations.

3.3.1 Cosmological constant and zero-point energy

As already noted, $\rho_{\rm V}$ given by (3.6) diverges in the ultraviolet: $\rho_{\rm V}$ behaves as k^4 for $k \to \infty$. However, there is the expectation that quantum field theory as well as general relativity break down at certain high-energy scale, perhaps, of the order of the reduced Planck energy $M_{\rm Planck} = 10^{18}$ GeV, above which they must be replaced by a more fundamental theory (MFT, for short) being still unknown.

This situation resembles classical electrodynamics that suffers from the ultraviolet catastrophe (see, for instance, [29]). The resolution of this problem eventually resulted in a development of quantum electrodynamics.

Analogously, it is usually presumed that quantum field theory arises as an effective low-energy theory from MFT. This may legitimate the cutoff of the integration in (3.6) at certain $|\mathbf{k}| = M_{\rm UV}$, so that it becomes finite[(6)]

$$\rho_{\rm V}^{\rm UV} = \frac{M_{\rm UV}^4}{16\pi^2}\left[\left(1 + \frac{m^2}{2M_{\rm UV}^2}\right)\sqrt{1 + \frac{m^2}{M_{\rm UV}^2}} - \frac{m^4}{2M_{\rm UV}^4}\ln\left(\frac{M_{\rm UV} + \sqrt{M_{\rm UV}^2 + m^2}}{m}\right)\right] . \tag{3.13}$$

Assuming $M_{\rm UV} \gg m$, one immediately obtains $\rho_{\rm V} \approx M_{\rm UV}^4/16\pi^2$. If we take that $M_{\rm UV} = M_{\rm Planck}$, then we roughly have

$$\rho_{\rm V}^{\rm Planck}/\rho_{\rm de} \approx 10^{121} , \tag{3.14}$$

[(6)]It seems that Nernst (1916) and Pauli (1920s) were the first who estimated this integral by making this cutoff (see [21, 48] and references therein). However, as it is argued in [49], a truncation of the high-energy modes at some finite physical energy scale is actually illegitimate, because certain effects of quantum field theory cannot be understood without taking into account holistic aspects of the theory.

i.e. an enormous discrepancy between this theoretical estimate and observations.

According to supersymmetric theory (SUSY) [50, 51], each boson has a fermionic superpartner with the same physical parameters and vice versa. The sum of the zero-point energies of such a pair turns out to be precisely zero. However, since the superpartners of the standard model particles have not been detected, supersymmetry must be broken to avoid conflict with observations. The energy scale M_{SUSY} below which the symmetry is definitely unobserved is of the order of 10^3 GeV – energy scale that is available in modern accelerators. Taking $M_{\mathrm{UV}} = M_{\mathrm{SUSY}}$, one still obtains unacceptable result:

$$\rho_{\mathrm{V}}^{\mathrm{SUSY}}/\rho_{\mathrm{de}} \approx 10^{61} . \tag{3.15}$$

The three-dimensional cutoff regularization, however, does not respect the local Lorentz invariance [52, 53, 54].(7) The violation of a symmetry by a regularization can lead to unphysical consequences. For instance, in QED, the order-α vacuum polarization diagram is also ultraviolet divergent. A regularization by imposing a four-dimensional Euclidean cutoff (see below) results in a nonzero photon mass proportional to the cutoff. This obviously contradicts reality. The reason is that this regularization violates the Ward identities and, hence, the gauge symmetry [8]. Another examples are mentioned in [53, 54].

To preserve the local Lorentz invariance, one may make the four-dimensional cutoff. This can be done by noting that ρ_{V} given in (3.6) can be rewritten as

$$\rho_{\mathrm{V}} = \frac{m^2}{4}\int \frac{d^4k}{(2\pi)^4}\frac{i}{k^2 - m^2 + i\varepsilon} = \frac{m^2}{4} D_{\mathrm{F}}(0) \tag{3.16}$$

(see Appendix B for details), where $D_{\mathrm{F}}(0)$ is the Feynman propagator [8] evaluated at $x = 0$, explicitly telling us that this is a one-loop vacuum energy (see Figure 3.1 (a)). Then performing a Wick rotation in (3.16) and imposing the four-momentum Euclidean cutoff ($k_0 \to ik_{\mathrm{E}0}$ and $-k^2 \to k_{\mathrm{E}}^2 = k_{\mathrm{E}0}^2 + |\mathbf{k}|^2 = M_{\mathrm{UV}}^2$ [8, 45]), we obtain

$$\rho_{\mathrm{V}} = \frac{m^2}{4}\int \frac{d^4k_{\mathrm{E}}}{(2\pi)^4}\frac{1}{k_{\mathrm{E}}^2 + m^2} = \frac{m^2}{64\pi^2}\left[M_{\mathrm{UV}}^2 - m^2 \ln\left(\frac{M_{\mathrm{UV}}^2 + m^2}{m^2}\right)\right], \tag{3.17}$$

i.e. the one-loop vacuum energy diverges as M_{UV}^2 [52, 53].

If we switch on non-gravitational interactions among the quantum fields, then multiple loops produce a more dramatic divergence when the cutoff M_{UV} goes to infinity. Indeed, applying the path integral formalism for obtaining vacuum Feynman diagrams in the case of $\lambda\phi^4$-theory, we easily derive

$$\begin{aligned} E_0 &= \frac{Zm_0^2}{4}\int d^4x\, D_{\mathrm{F}}(x-x) + \frac{Z^2\lambda_0}{8}\int d^4x\, D_{\mathrm{F}}^2(x-x) \\ &\quad -\frac{iZ^4\lambda_0^2}{48}\int d^4x d^4y\Big(3D_{\mathrm{F}}(x-x)D_{\mathrm{F}}(y-y)D_{\mathrm{F}}^2(x-y) + D_{\mathrm{F}}^4(x-y)\Big) + \mathrm{O}(Z^6\lambda_0^3), \end{aligned} \tag{3.18}$$

where the second and third terms are due to the self-interaction of the scalar field and are diagrammatically presented in Figure 3.1 (b) and (c) with (d). The Feynman propagator in (3.18) is given by

$$D_{\mathrm{F}}(x-y) = \int \frac{d^4k}{(2\pi)^4}\frac{ie^{-ik(x-y)}}{Zk^2 - Zm_0^2 + i\varepsilon}, \tag{3.19}$$

(7)For experimental tests of the local Lorentz invariance, see a review [55].

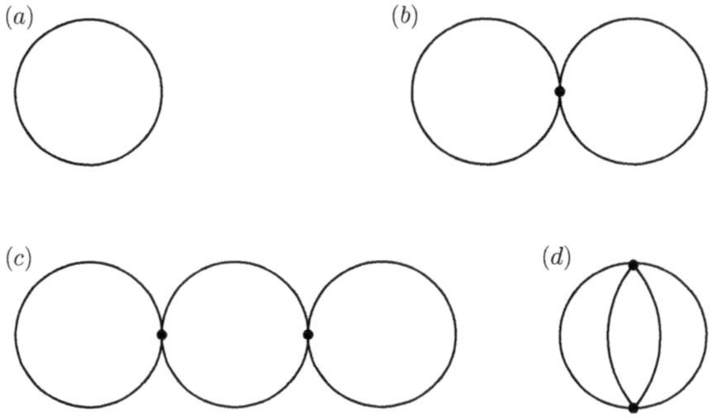

Figure 3.1: Vacuum diagrams up to the order-λ^2 in $\lambda\phi^4$-theory.

where m_0, λ_0 and Z are bare mass, coupling constant and field-strength renormalization:

$$\begin{aligned} Z &\equiv 1 + \delta_{\rm Z}^{(2)}\lambda^2 + {\rm O}(\lambda^3)\,, \\ Zm_0^2 &\equiv m^2 + \delta_{\rm m}^{(1)}\lambda + \delta_{\rm m}^{(2)}\lambda^2 + {\rm O}(\lambda^3)\,, \\ Z^2\lambda_0 &\equiv \lambda + \delta_{\lambda}^{(2)}\lambda^2 + {\rm O}(\lambda^3)\,. \end{aligned} \tag{3.20}$$

Here, m and λ are the physical mass and coupling constant, respectively, fixed by the renormalization conditions [8, 28, 56].

Expanding E_0 as a series in λ and dividing the result by the volume of the four-dimensional manifold, we obtain its energy density in the following form

$$\begin{aligned} \rho_{\rm V} &= \frac{m^2}{4}J_{1,4} + \frac{m^2\lambda}{8}J_{1,4}J_{2,4} + \frac{\lambda^2}{48}\Big(6\,\delta_{\lambda}^{(2)}J_{1,4}^2 - \mathcal{I} \\ &\quad + 12\big(\delta_{\rm m}^{(2)} - m^2\delta_{\rm Z}^{(2)}\big)\big(J_{1,4} - m^2 J_{2,4}\big) + 3m^2 J_{1,4}^2 J_{3,4}\Big) + {\rm O}(\lambda^3)\,, \end{aligned} \tag{3.21}$$

where $J_{\alpha,4}$ is given in (7) and $\mathcal{I}$ is defined as[8]

$$\mathcal{I} = \int \frac{d^4k_{\rm E}}{(2\pi)^4}\frac{d^4p_{\rm E}}{(2\pi)^4}\frac{d^4q_{\rm E}}{(2\pi)^4}\,\frac{1}{k_{\rm E}^2+m^2}\,\frac{1}{p_{\rm E}^2+m^2}\,\frac{1}{q_{\rm E}^2+m^2}\,\frac{1}{(k_{\rm E}+p_{\rm E}+q_{\rm E})^2+m^2}\,. \tag{3.22}$$

In deriving (3.21), we have used an equality $\delta_{\rm m}^{(1)} = -\frac{1}{2}J_{1,4}$ calculated in [8].

Taking into account that $J_{1,4} \sim M_{\rm UV}^2$, $J_{2,4} \sim \ln(M_{\rm UV}/m)$, $J_{3,4} \sim M_{\rm UV}^{-2}$ as well as from dimensional considerations $\delta_{\rm Z}^{(2)} \sim \ln(M_{\rm UV}/m)$, $\delta_{\rm m}^{(2)} \sim M_{\rm UV}^2\ln(M_{\rm UV}/m)$, $\delta_{\lambda}^{(2)} \sim \ln(M_{\rm UV}/m)$ [8, 58], and $\mathcal{I} \sim M_{\rm UV}^4\ln(M_{\rm UV}/m)$, one obtains

$$\rho_{\rm V} \sim \lambda^2 M_{\rm UV}^4 \ln(M_{\rm UV}/m)\,. \tag{3.23}$$

---

(8) For an analytic expression of this integral, see an article [57].

Note that the order-λ vacuum energy diverges "merely" as $\lambda M_{\rm UV}^2 \ln(M_{\rm UV}/m)$ [53].

Since $M_{\rm UV}^2$ is defined as $k_{\rm E}^2 = k_{\rm E0}^2 + |\mathbf{k}|^2 = -\eta^{\mu\nu} k_\mu k_\nu$, we are not allowed to ascribe physical meaning to $M_{\rm UV}$ as a finite energy scale below which the theory is reliable. Moreover, the first term in (3.18) up to the volume of the manifold comes from

$$\frac{1}{2} \int \frac{d^4 k_{\rm E}}{(2\pi)^4} \ln\left(\frac{k_{\rm E}^2 + m_0^2}{\mu^2}\right) \tag{3.24}$$

(see Appendix B for details). It is quite obvious that if we regularize this integral by imposing the four-dimensional Euclidean cutoff, we find that it diverges as $M_{\rm UV}^4 \ln(M_{\rm UV}/\mu)$. This does not coincide with (3.17).

Dimensional regularization, as noted in the previous section, gives the same results for both integrals and introduces the mass scale μ after $\overline{MS}$ renormalization [8, 45] (see also Appendix A).

To our knowledge, the latest naive estimate of the zero-point energy density contributing to the cosmological constant we could find in the literature is made in [54] and equals $\rho_{\rm V} \approx -10^9$ GeV^4 in the one-loop approximation. This vacuum energy is related to the heaviest standard model particles, where μ is taken to be the geometric mean of photon ($\lambda \approx 500$ nm) and graviton ($\lambda \approx 10^{26}$ m) energies. This value is still physically unacceptable.

Before we proceed, we note that it is inconsistent not to take into account the bare cosmological constant Λ_0 in (3.12) which value is dictated by MFT. As mentioned above, the vacuum energy density $\rho_{\rm V}$ appearing in (3.12) is not observable, but Λ is. Thus, one may consider the large theoretical values of $\rho_{\rm V}$ as an evidence that we actually need to develop MFT.

3.3.2 Cosmological constant and spontaneous symmetry breaking

There are other sources of vacuum energy contributing to the cosmological constant. These are associated with spontaneous symmetry breaking (SSB) and results from nonzero vacuum expectation values of certain fields.

Specifically, in the $SU(5)$ theory [59], being the most simple grand unified theory, after the first phase transition associated with the spontaneous symmetry breaking $SU(5) \to SU(3)_{\rm C} \times SU(2)_{\rm L} \times U(1)_{\rm Y}$ at energy scale $M_{\rm GUT} = 10^{16}$ GeV, the second one $SU(3)_{\rm C} \times SU(2)_{\rm L} \times U(1)_{\rm Y} \to SU(3)_{\rm C} \times U(1)_{\rm EM}$ occurs at $M_{\rm EW} = 10^2$ GeV followed by a spontaneous breaking of chiral symmetry at $M_{\rm QCD} = 0.3$ GeV: $U(N_f)_{\rm R} \times U(N_f)_{\rm L} \to SU(N_f)_{\rm V}$, where N_f is the number of light quark flavors [60, 61, 62]. Let us consider the last two SSBs in more detail.

Electroweak symmetry breaking According to the electroweak theory proposed by Glashow, Weinberg and Salam that is based on a semisimple gauge group $SU(2)_{\rm L} \times U(1)_{\rm Y}$ (each group has its own coupling constant, see below), there is a fundamental scalar φ that is a doublet with respect to $SU(2)$ and possesses a weak hypercharge $Y_\varphi = +1/2$.

The part of the Lagrangian depending only on φ is

$$\mathcal{L}_\varphi = \frac{1}{2}(D_\mu\varphi)^\dagger(D^\mu\varphi) - V(\varphi^\dagger\varphi), \tag{3.25}$$

where $D_\mu = \partial_\mu - \frac{i}{2}gA^a_\mu\sigma^a - \frac{i}{2}g'B_\mu$ is the gauge covariant derivative, σ^a ($a = 1, 2, 3$) are the three Pauli matrices that are the generators of $SU(2)$, A^a_μ are the three gauge fields corresponding to $SU(2)$ and B_μ is the gauge field associated with $U(1)$.

The potential $V(\varphi^\dagger\varphi)$ is given by

$$V(\varphi^\dagger\varphi) = V_0 - \frac{\mu^2}{2}(\varphi^\dagger\varphi) + \frac{\lambda}{4}(\varphi^\dagger\varphi)^2, \tag{3.26}$$

where $\mu^2 > 0$ and $\lambda > 0$. For this particular potential a state with $\varphi = 0$ is not stable (because it is a local maximum of V), but $\varphi^\dagger\varphi = \mu^2/\lambda \equiv v^2/2$ is.

Now let us choose $A^a_\mu = 0$, $B_\mu = 0$ and $\langle\varphi\rangle = (0, v)^{\mathrm{T}}/\sqrt{2}$ as a vacuum state and consider (in the unitary gauge)

$$\varphi^{\mathrm{T}} = (0, v + H)/\sqrt{2}, \tag{3.27}$$

where H is a Hermitian field known as the Higgs boson. The Lagrangian rewritten via H is still invariant under the whole gauge group $SU(2)_{\mathrm{L}} \times U(1)_{\mathrm{Y}}$, but the nontrivial vacuum state chosen above is invariant under $U(1)_{\mathrm{EM}}$ ($\ni \exp(i\alpha(x)(\sigma^3 + 1)/2)$). In other words, $SU(2)_{\mathrm{L}} \times U(1)_{\mathrm{Y}}$ is spontaneously broken down to $U(1)_{\mathrm{EM}}$ [7, 8, 63].

Let us discuss the vacuum energy density resulting from $\langle\varphi\rangle \neq 0$. It is straightforward to derive, that for this particular configuration of the scalar field, its energy-momentum tensor takes the form

$$T_{\mu\nu}(\langle\varphi\rangle) = V(\langle\varphi\rangle)g_{\mu\nu} = \left(V_0 - \frac{m_{\mathrm{H}}^2}{8\sqrt{2}G_{\mathrm{F}}}\right)g_{\mu\nu}, \tag{3.28}$$

where $m_{\mathrm{H}}^2 \equiv \mu^2$ is the mass of the Higgs boson and $G_{\mathrm{F}} = 1/\sqrt{2}v^2$ is the Fermi constant equaling $1.16637(1) \times 10^{-5}$ GeV^{-2} [18].

According to preliminary results from CMS [64] and ATLAS [65], if the Higgs boson exists, then $m_{\mathrm{H}} \approx 125$ GeV, so that we find

$$\frac{m_{\mathrm{H}}^2}{8\sqrt{2}G_{\mathrm{F}}} \approx M_{\mathrm{EW}}^4 = 10^8\ \mathrm{GeV}^4. \tag{3.29}$$

If one puts $V_0 = 0$ in (3.26), then the Higgs condensate $\langle\varphi\rangle$ contributes a negative vacuum energy of the order of M_{EW}^4 to the cosmological constant. However, one may equally put $V(\langle\varphi\rangle) = 0$, so that, after the electroweak SSB, there is no contribution of the Higgs condensate to the total vacuum energy, but then we must admit enormous fine-tuning [66, 67, 68].

We merely note that the cosmological constant problem is accompanied by a similar puzzle in the standard model of particle physics, that is known as the hierarchy problem [69, 70]. Its essence consists in that any field, which is coupled to the Higgs boson, makes a contribution to m_{H}^2 that diverges as M_{UV}^2.

Chiral symmetry breaking and QCD condensates The chiral symmetry breaking is an example of SSB of a global symmetry $U(N_f)_{\mathrm{R}} \times U(N_f)_{\mathrm{L}}$, where N_f is the number of light flavors.

Let us consider the light fermionic part of the QCD Lagrangian[(9)]

$$\mathcal{L} = \bar{u}(i\gamma^\mu D_\mu - m_{\mathrm{u}})u + \bar{d}(i\gamma^\mu D_\mu - m_{\mathrm{d}})d\,, \tag{3.30}$$

where γ^μ are the four Dirac gamma matrices. The covariant derivative is given by $D_\mu = \partial_\mu - \frac{i}{2} g_s G^a_\mu \lambda^a$, where λ^a ($a = 1, \ldots, 8$) are the Gell-Mann matrices being the generators of $SU(3)$, g_s is a coupling constant associated with the local $SU(3)_{\mathrm{C}}$ and $G^a_{\mu\nu}$ are its eight gauge fields known as gluons.

If we put $m_{\mathrm{u}} = m_{\mathrm{d}} = 0$, then this Lagrangian is invariant under $U(2)_{\mathrm{R}} \times U(2)_{\mathrm{L}}$. However, m_{u} and m_{d} are small, but nonzero in reality. It was hypothesized that the chiral symmetry is spontaneously broken by a quark condensate $\langle \bar{q}q \rangle$ down to the diagonal $SU(2)$ part of the chiral group $U(2)_{\mathrm{R}} \times U(2)_{\mathrm{L}}$, where pions $\pi^\pm$ and π^0 (as being light) correspond to three pseudo-Goldstone bosons [8, 71, 72].

The quark condensate $\langle \bar{q}q \rangle = \langle \bar{u}u + \bar{d}d \rangle$ together with the gluon condensate $\langle G^a_{\mu\nu} G^{\mu\nu a} \rangle$ are constituents of the QCD vacuum that is non-perturbative, since g_s becomes larger at low energies, however smaller at high energies – a phenomenon known as asymptotic freedom [73]. This makes a discussion of the QCD vacuum to be a complicated issue.

One usually identifies $\langle \bar{q}q \rangle \sim -M^3_{\mathrm{QCD}}$ and $\langle G^a_{\mu\nu} G^{\mu\nu a} \rangle \sim M^4_{\mathrm{QCD}}$ at energy scale corresponding to $g_s \sim 1$ [72, 74]. These condensates are conventionally considered to be properties of the QCD vacuum and to be constant throughout spacetime,[(10)] thus they contribute an energy density of roughly 10^{-2} GeV^4 to the cosmological constant, which is approximately 10^{45} orders of magnitude larger than $\rho_{\mathrm{de}} \approx 2.9 \times 10^{-47}$ GeV^4.

---

(9) The s-quark is also light with respect to the QCD energy scale M_{QCD}, but we do not consider it here for the sake of making our discussion as transparent as possible.

(10) Note, there exists another viewpoint on it, namely that the QCD condensates are localized within hadrons only, so that the contribution of the QCD vacuum is already taken into account in the hadrons masses. For more details, see [75] and references therein as well as [62]. For criticism of this statement, see [76].

Chapter 4

Particular approaches to CCP

Many approaches of how to solve the cosmological constant problem have been suggested in the literature. A survey of these different proposals can be found in [14, 21, 62, 68, 77, 78] as well as [79], where they are classified into five classes: fine-tuning, approaches based on a certain symmetry, backreaction mechanisms, violation of the equivalence principle and statistical approaches. However, the problem remains unsolved.

In the present chapter, we shall briefly discuss some of the ideas being relevant for our further discussions and introduce so-called q-theory [4, 80, 81].

4.1 Fine-tuning adjustment

Let us consider a real scalar field ϕ governed by the following Lagrangian

$$\mathcal{L}_\phi \;=\; -\nabla_\mu\phi\nabla^\mu\phi + V(\phi)\,. \tag{4.1}$$

Setting the variation of $\mathcal{L}_\phi$ over ϕ to zero yields

$$\nabla^2\phi + V'(\phi) \;=\; 0\,, \tag{4.2}$$

where $V' \equiv dV/d\phi$. The scalar energy-momentum tensor $T_{\mu\nu}(\phi)$ is

$$T_{\mu\nu}(\phi) \;=\; 2\nabla_\mu\phi\nabla_\nu\phi - g_{\mu\nu}\big(\nabla_\lambda\phi\nabla^\lambda\phi - V(\phi)\big)\,. \tag{4.3}$$

A constant configuration $\phi = \phi_0 \neq 0$, such that

$$V'|_{\phi=\phi_0} \;=\; 0 \tag{4.4}$$

reduces $T_{\mu\nu}(\phi)$ to $V(\phi_0)g_{\mu\nu}$. In other words, $\rho_\phi = -P_\phi = V(\phi_0)$. Therefore, taking a proper potential $V(\phi)$, one can always bring the total vacuum energy in the Einstein equations into the physically acceptable level.

This approach implies an artificial choice of the potential $V(\phi)$, such as the observed value of dark energy ρ_{de} equals $V(\phi_0)$ plus large negative number coming from quantum fields. This makes it unattractive.

4.2 Dynamical adjustment

Dolgov in 1983 [77] and Ford in 1987 [82] proposed a model that provides a dynamical cancellation of a cosmological constant by a massless scalar field non-minimally coupled with gravity.

This model is determined by the following Lagrangian density

$$\mathcal{L}_\phi \;=\; -\nabla_\mu\phi\nabla^\mu\phi + \zeta_R R\phi^2 \,. \tag{4.5}$$

The scalar field equation is given by

$$\left(\nabla^2 + \zeta_R R\right)\phi \;=\; 0 \,, \tag{4.6}$$

while its energy-momentum tensor is found to be

$$\begin{aligned} T_{\mu\nu}(\phi) \;&=\; 2\nabla_\mu\phi\nabla_\nu\phi - g_{\mu\nu}\nabla_\lambda\phi\nabla^\lambda\phi \\ &\quad + \zeta_R\left(R\phi^2 g_{\mu\nu} - 2R_{\mu\nu}\phi^2 + 2\nabla_\mu\nabla_\nu\phi^2 - 2g_{\mu\nu}\nabla^2\phi^2\right) . \end{aligned} \tag{4.7}$$

It is straightforward to show that there exist at least two cosmological solutions of the scalar and Einstein field equations. The first solution corresponds to de Sitter spacetime

$$\phi(t) \;=\; 0, \quad H(t) \;=\; \sqrt{\Lambda/3}\,, \quad \rho_{\rm m}(t) \;=\; 0\,, \tag{4.8}$$

where $\rho_{\rm m}(t)$ is the energy density of the matter field. The second solution is

$$\phi(t) \;=\; \pm\left(\frac{8\zeta_R\rho_\Lambda}{3 + 28\zeta_R + 60\zeta_R^2}\right)^{1/2} \times t\,, \quad H(t) \;=\; \frac{1}{2t}\left(1 + \frac{1}{2\zeta_R}\right) \tag{4.9}$$

with

$$\rho_{\rm m}(t) \;=\; \frac{3}{32\pi G t^2}\left(1 + \frac{1}{2\zeta_R}\right)^2 . \tag{4.10}$$

Considering small homogeneous perturbations around these solutions and linearizing the scalar, Einstein and matter field equations with respect to them, one easily finds that (4.8) is asymptotically stable only if $\zeta_R < 0$ and $P_{\rm m}/\rho_{\rm m} \equiv w_{\rm m} > -1$, while $\phi(t)$ and $H(t)$ given in (4.9) are asymptotically stable only if

$$\zeta_R \;\in\; D, \quad \text{where} \quad D \;=\; \big(-\infty, -1/2\big) \cup \big(0, +\infty\big)\,, \tag{4.11}$$

but $t^2\rho_{\rm m}(t)$ approaches a constant for large time, which depends on the initial conditions imposed. Thus the cosmological constant Λ for $\zeta_R \in D$ is dynamically compensated by the scalar field subject to the condition that

$$\zeta_R\Lambda(3 + 28\zeta_R + 60\zeta_R^2) \;>\; 0 \tag{4.12}$$

is satisfied for ϕ to be real.

However, this model is unrealistic [77, 82]. Indeed, considering the weak-field limit, one finds that the effective gravitational constant is

$$G_{\text{eff}} = \frac{G}{1 + 16\pi G \zeta_R \phi^2} \sim 1/t^2 \tag{4.13}$$

at large time. As a consequence, $\dot{G}_{\text{eff}}/G_{\text{eff}} \approx -2H_0 \approx -10^{-10}$ yr^{-1}. This contradicts observations [83, 84] (see also [55]).

4.3 Q-theory

As already mentioned, there is the expectation that general relativity and quantum field theory are low-energy effective theories. This means that they must be replaced by a more fundamental theory that works even at high-energy scales and simplifies to general relativity and quantum field theory in the limits $\hbar \to 0$ and $G \to 0$, respectively. Quantum gravity is, however, not yet established. Q-theory is a phenomenological approach to the quantum vacuum [4].

According to q-theory, the vacuum is a Lorentz invariant self-sustained medium, that is characterized by a conserved relativistic scalar q. This parameter describes microscopic, high-energy degrees of freedom. On the other hand, the q-variable allows to discuss macroscopic, low-energy physics, because it obeys the conservation law (see below). Condensed-matter analog of the q-variable is the particle density n in liquids.

The microscopic vacuum energy $\epsilon_{\text{micro}}(q)$ could be of the order of M_{UV}^4. The macroscopic vacuum energy $\epsilon_{\text{macro}}(q)$, that appears in the equations of the low-energy effective theories, can be of the order of the observed cosmological constant. However, since vacuum is self-sustained, $\epsilon_{\text{macro}}(q_0) = 0$, where $q_0 \neq 0$ is a value of q in the equilibrium state. Thus, the small value of ρ_{de} is due to the fact that $q_{\text{today}} \approx q_0$.

Let us consider several examples of the q-variable which will clarify what was just said.

First example The q-variable can be realized by a tensor of the third rank $A_{\mu\nu\lambda}$, namely

$$q \propto e^{\mu\nu\lambda\rho} \nabla_\mu A_{\nu\lambda\rho} , \tag{4.14}$$

where $e_{\mu\nu\lambda\rho}$ is the absolutely antisymmetric tensor: $e_{\mu\nu\lambda\rho} \equiv \sqrt{-g}\, \varepsilon_{\mu\nu\lambda\rho}$, where $\varepsilon_{\mu\nu\lambda\rho}$ is the Levi-Civita symbol [4, 85, 86, 87, 88].

Let us suppose the dynamics of the q-variable is governed by

$$S[g, A] = -\int d^4x \sqrt{-g}\, \epsilon_{\text{micro}}(q) , \tag{4.15}$$

where $\epsilon_{\text{micro}}(q)$ is a generic function of its variable. The q-variable depends on both $A_{\mu\nu\lambda}$ and $g_{\mu\nu}$. The variation of (4.15) with respect to $A_{\mu\nu\lambda}$ gives

$$\nabla_\mu \left(\epsilon'_{\text{micro}}\right) = 0 \;\Rightarrow\; \epsilon'_{\text{micro}} = \mu = \text{const} , \tag{4.16}$$

where prime stands for derivation with respect to the q-variable. This equation represents the conservation law of q. The variation of $S[g, A]$ with respect to the metric gives us its energy-momentum tensor

$$T_{\mu\nu}(q) \;=\; \epsilon_{\text{macro}}(q) g_{\mu\nu}\,, \quad \text{where} \quad \epsilon_{\text{macro}}(q) \;\equiv\; \epsilon_{\text{micro}}(q) - q\,\epsilon'_{\text{micro}}(q)\,. \tag{4.17}$$

Thus, the vacuum energy density that enters the Einstein field equations is given by $\epsilon_{\text{macro}}(q)$, rather than $\epsilon_{\text{micro}}(q)$.

As just mentioned, $q_{\text{today}} \approx q_0$ and, hence,

$$\rho_{\text{de}} \;=\; \epsilon_{\text{macro}}(q)|_{q_{\text{today}}} \;=\; \Big(\epsilon_{\text{micro}}(q) - q\,\epsilon'_{\text{micro}}(q)\Big)\Big|_{q_{\text{today}}} \;\ll\; M^4_{\text{UV}}, \tag{4.18}$$

while both $\epsilon_{\text{micro}}(q_0)$ and $q_0\,\epsilon'_{\text{micro}}(q_0)$ could be of the order of M^4_{UV}.[(1)]

Note that a fundamental scalar field cannot be regarded as the q-variable. The fact is that for the scalar field μ is identically zero as it follows from (4.4), while μ is determined by the equilibrium state of the vacuum and, generally speaking, nonzero [6]. Consequently, q is not a fundamental scalar, i.e. it must be built out of some fundamental tensor fields and their covariant derivatives for having dynamics.

Second example The second example of such kind of the variable can be realized by the following pseudoscalar

$$q \;\propto\; F_{\mu\nu}\widetilde{F}^{\mu\nu}, \tag{4.19}$$

where $F_{\mu\nu} \equiv \nabla_\mu A_\nu - \nabla_\nu A_\mu$ is the field strength tensor, $\widetilde{F}_{\mu\nu}$ is a dual tensor. Indeed, one has a sequence of equalities

$$F_{\mu\nu}\widetilde{F}^{\mu\nu} \;=\; \frac{1}{2}\,e^{\mu\nu\lambda\rho}F_{\mu\nu}F_{\lambda\rho} \;=\; e^{\mu\nu\lambda\rho}F_{\mu\nu}\nabla_\lambda A_\rho \;=\; e^{\mu\nu\lambda\rho}\nabla_\lambda\big(F_{\mu\nu}A_\rho\big)\,. \tag{4.20}$$

Since $F_{\mu\nu}A_\rho$ does not depend on the metric field, it can be associated with $A_{\mu\nu\rho}$ from the previous example. It is also worth mentioning that a vector dual to $F_{\mu\nu}A_\rho$ is proportional to the topological or Chern-Simons current [89].

For a gauge $SU(N)$ vector field, one can define the q-variable in a completely analogous manner:

$$q \;\propto\; \text{tr}\big(F_{\mu\nu}\widetilde{F}^{\mu\nu}\big)\,, \quad \text{where} \quad F_{\mu\nu} \;\propto\; [D_\mu, D_\nu]\,. \tag{4.21}$$

Here D_μ is the covariant derivative associated with the local $SU(N)$. We note that the q-variable related with the gluonic vacuum was discussed in [90].

[(1)] A particular model with the q-variable was proposed in [85]. Two extra equations, so-called equilibrium conditions, were used there for getting Minkowski spacetime ($H = \dot{a}/a = 0$) in the equilibrium state of vacuum (see Eq. (3.4) in [85]). However, at it is argued in [81], this corresponds to a fine-tuning of the integration constant μ in (4.16). So instead of the fine-tuning of the action, there must be done the fine-tuning of initial conditions. We will see below, that, in principle, this particular problem can be overcome.

Third example Another comparatively simple example of the q-variable is given by

$$q \;\propto\; \nabla^\mu A_\mu\,, \tag{4.22}$$

with the Lagrangian density as in (4.15).

If one adds extra terms depending on the vector field A_μ to this Lagrangian, then the simple form of both the conservation law (4.16) and the energy-momentum tensor (4.17) are spoiled. Nevertheless, it can be that the q-variable appears asymptotically, i.e. in the limit $t \to \infty$.

Long ago Dolgov proposed a vector model with the dynamical compensation of a cosmological constant [5]. This model is governed by the following Lagrangian density

$$\mathcal{L}_\mathrm{A} \;=\; \zeta_1(\nabla_\mu A_\nu)(\nabla^\mu A^\nu)\,. \tag{4.23}$$

Taking $A_\mu = A_0(t)\delta^0_\mu$, he found a solution behaving as

$$A_0(t) \;=\; \pm\sqrt{\rho_\Lambda/4\zeta_1}\times t\,, \quad H(t) \;=\; 1/t\,. \tag{4.24}$$

This solution is asymptotically stable, i.e. does not depend on initial conditions imposed on dynamical variables and provides with the exact dynamical cancellation of the vacuum energy in the Einstein equations. The q-variable appears here as [6]

$$A_{\mu;\nu}(t) \;=\; (\delta^0_\mu\partial_\nu - \Gamma^0{}_{\mu\nu})A_0(t) \;\to\; q g_{\mu\nu}, \quad \text{when} \quad t \;\to\; \infty\,, \tag{4.25}$$

where $q \equiv \pm\sqrt{\rho_\Lambda/4\zeta_1}$.

However, Dolgov's model has at least two obstacles making it unrealistic. First, the Hubble parameter $H = 1/t$ implies $a(t) \sim t$. This formally corresponds to the string-dominated universe and is in an obvious contradiction with the observational data. Second, Newton's law of gravity is violated as well as the properties of gravitational waves in comparison with them in general relativity [91].

Inspired by q-theory approach and Dolgov's model, we will generalize (4.23) in the next chapter by adding extra terms quadratically depending on the vector field. We will, however, refrain from the reference to q-theory in the next two chapters. The reason for that is that our further analysis is independent of it and may be regarded as a search of a realistic model with a dynamical cancellation of the total vacuum energy.

Chapter 5

Vector-tensor model I

Let us consider a hypothetical universe governed by the following effective action

$$S[g,\psi,A] = -\int d^4x\sqrt{-g}\left(\frac{1}{2}M^2_{\text{Planck}}\Big(R+2\Lambda\Big)+\mathcal{L}_{\text{A}}+\mathcal{L}_{\text{m}}\right), \tag{5.1}$$

where M_{Planck} is the reduced Planck mass defined in terms of the gravitational constant G, $\mathcal{L}_{\text{m}}$ is the Lagrangian of matter field ψ and $\mathcal{L}_{\text{A}}$ is the Lagrangian of a vector field A_μ non-minimally coupled with gravity

$$\mathcal{L}_{\text{A}} = \zeta_1(\nabla_\mu A_\nu)(\nabla^\mu A^\nu)+\zeta_2(\nabla_\nu A_\mu)(\nabla^\mu A^\nu)+\zeta_3(\nabla^\mu A_\mu)^2+\zeta_4 R A_\mu A^\mu\,. \tag{5.2}$$

This vector Lagrangian (5.2) was proposed long ago as one of the alternative metric theories of gravity [92, 93, 94]. In general, vector-tensor theories may be divided into two subclasses [55]: Einstein-aether theories in which A_μ is the unit timelike four-vector [95] and unconstrained vector-tensor theories corresponding to (5.2). The purpose of the former models is to introduce a dynamical violation of the local Lorentz invariance by means of the appearance of a preferred rest frame. This is motivated by purely theoretical evidences that this symmetry might be broken at high-energy scales [96]. As concerns the latter, a model with (5.1) and (5.2), where Λ is precisely zero, has been recently considered in [97]. The vector field is invoked there to describe both dark matter and dark energy in our universe.

However, the nonzero cosmological constant influences a behavior of the vector field A_μ, such that Λ is compensated by A_μ at late times for certain choices of the ζ-coefficients in (5.2) [5, 98]. It will be proven that independent of how large Λ is, the vector field does the job for a large domain of initial conditions imposed on dynamical variables. In addition, it will be shown that short-wavelength perturbation of the vector and metric fields gives an unacceptable modification of the Newton gravity law in a linear approximation. Thus, this hypothetical universe has nothing to do with our own universe. In the next chapter, we will consider much more sophisticated vector-tensor model in order to avoid this difficulty. The analysis made below will be helpful there.

For the sake of convenience, we rewrite (5.2) via symmetric and skew-symmetric tensors built out of the first covariant derivative of the vector field $\nabla_\mu A_\nu$, namely

$$S_{\mu\nu} \equiv \nabla_\mu A_\nu + \nabla_\nu A_\mu , \quad F_{\mu\nu} \equiv \nabla_\mu A_\nu - \nabla_\nu A_\mu . \tag{5.3}$$

The half sum of $S_{\mu\nu}$ and $F_{\mu\nu}$ gives us the first term in (5.2) and the half difference of them gives the second term of the Lagrangian $\mathcal{L}_\mathrm{A}$, so that we can rewrite (5.2) as

$$\mathcal{L}_\mathrm{A} = \frac{1}{4}\zeta_S\, S_{\mu\nu}S^{\mu\nu} + \frac{1}{4}\zeta_F\, F_{\mu\nu}F^{\mu\nu} + \frac{1}{4}\zeta_Q\,(S^\mu_\mu)^2 + \zeta_R R A_\mu A^\mu , \tag{5.4}$$

where we have used the equality $S^\mu_\mu = 2\nabla^\mu A_\mu$ resulting from (5.3) and defined new coefficients as

$$\zeta_S \equiv \zeta_1 + \zeta_2 , \quad \zeta_F \equiv \zeta_1 - \zeta_2 , \quad \zeta_Q \equiv \zeta_3 , \quad \zeta_R \equiv \zeta_4 \tag{5.5}$$

which will be used in the following.

In principle, we could also add to the vector Lagrangian a quadratic term in the vector field A_μ, constructed from the contraction of the vector field with the Ricci tensor $R_{\mu\nu}$. However, this term does not give a new contribution to the model, because of the following correspondence

$$R_{\mu\nu}A^\mu A^\nu \longleftrightarrow -\frac{1}{4}S_{\mu\nu}S^{\mu\nu} + \frac{1}{4}F_{\mu\nu}F^{\mu\nu} + \frac{1}{4}(S^\mu_\mu)^2 , \tag{5.6}$$

which can be straightforwardly verified. Note, this correspondence (5.6) enables us to eliminate $S_{\mu\nu}S^{\mu\nu}$ from the Lagrangian, i.e. we can also write

$$\mathcal{L}_\mathrm{A} = \frac{1}{2}(\zeta_S + \zeta_F)F_{\mu\nu}F^{\mu\nu} + \frac{1}{4}(\zeta_S + \zeta_Q)(S^\mu_\mu)^2 - \zeta_S R_{\mu\nu}A^\mu A^\nu + \zeta_R R A_\mu A^\mu , \tag{5.7}$$

so that in the case of Minkowski spacetime $g_{\mu\nu} = \eta_{\mu\nu}$, the last two terms vanish and we arrive to the well-known vector model with a term fixing a gauge.

5.1 Vector and Einstein field equations

The vector field equation can be derived by calculating the functional derivative of the action (5.1) with respect to the field A_μ and setting it equal to zero. This yields

$$\zeta_S\, S^\lambda_{\mu;\lambda} + \zeta_F\, F^\lambda{}_{\mu;\lambda} + \zeta_Q\, S^\lambda_{\lambda;\mu} = 2\zeta_R\, R A_\mu . \tag{5.8}$$

Varying the action (5.1) with respect to the metric field $g_{\mu\nu}$, we obtain the Einstein field equations

$$R_{\mu\nu} - \frac{1}{2}R g_{\mu\nu} = M^{-2}_{\text{Planck}}\Big(\rho_\Lambda g_{\mu\nu} + T^\mathrm{A}_{\mu\nu} + T^\mathrm{m}_{\mu\nu}\Big) , \tag{5.9}$$

where $T^\mathrm{m}_{\mu\nu}$ is the energy-momentum tensor of the matter field ψ which is assumed to have the form (2.10), $T^\mathrm{A}_{\mu\nu}$ is the energy-momentum tensor of the vector field which we split into four terms

$$T^\mathrm{A}_{\mu\nu} = T^S_{\mu\nu} + T^F_{\mu\nu} + T^Q_{\mu\nu} + T^R_{\mu\nu} , \tag{5.10}$$

where each of $T^f_{\mu\nu}$, $f \in \{S, F, Q, R\}$ are presented in Appendix C.

5.2 Flat, homogeneous and isotropic universe

The flat, homogeneous and isotropic universe is described by the Friedmann-Robertson-Walker metric (2.4) with the zero curvature constant ($k = 0$), namely

$$ds^2 \; = \; g_{\mu\nu}dx^\mu dx^\nu \; = \; dt^2 - a^2(t)d\mathbf{r}^2 \,. \tag{5.11}$$

The vector field must be homogeneous as well, i.e.

$$A_\mu(x) \; = \; \big(A_0(t),\, A_i(t)\big)\,. \tag{5.12}$$

Generally speaking, (5.12) does not imply that the vector EMT is compatible with the isotropic universe. For that to be, $T^{\rm A}_{\mu\nu}$ must take the following form: $T^{\rm A}_{0i} = 0$ and $T^{\rm A}_{ij} \propto g_{ij}$ (see Section 2.2). However, it turns out that $T^{\rm A}_{0i}$ always vanish, but non-diagonal elements of $T^{\rm A}_{ij}$ definitely disappear only if $A_i(t) = 0$. This does not mean that there is no nontrivial $A_i(t)$, such that $T^{\rm A}_{ij} \propto g_{ij}$ as we will show below.

In the present section, we will proceed as follows: we firstly consider the case $A_\mu = (A_0,\, 0)$, then we will move on $A_\mu = (0,\, A_i)$, where $A_i(t)$ is a particular function of cosmic time t, such that non-diagonal elements of the energy-momentum tensor $T^{\rm A}_{\mu\nu}$ vanish and, finally, we will treat $A_\mu = (A_0,\, A_i)$.

5.2.1 Case: $A_\mu = (A_0, 0)$

Taking into account (5.11) and (5.12) with $A_i(t) = 0$, one derives from (5.8) and (5.9) together with (5.10) that

$$\ddot{v} + 3h\dot{v} + 3\Big(\dot{h} + \zeta^{-1}\big[(2\zeta_R - \zeta_S)\dot{h} + (4\zeta_R - \zeta_S)h^2\big]\Big)v \; = \; 0\,, \tag{5.13a}$$

$$3h^2 - \lambda - r_v - r_{\rm m} \; = \; 0\,, \tag{5.13b}$$

$$2\dot{h} + 3h^2 - \lambda + p_v + w_{\rm m}r_{\rm m} \; = \; 0\,, \tag{5.13c}$$

$$\dot{r}_{\rm m} + 3h(1 + w_{\rm m})r_{\rm m} \; = \; 0\,, \tag{5.13d}$$

where by definition

$$\zeta \; \equiv \; \zeta_Q + \zeta_S \; \neq \; 0 \tag{5.14}$$

and, for the sake of clarity, dimensionless variables have been introduced by rescaling the original ones with appropriate powers of the reduced Planck mass, namely

$$\begin{aligned} v \;&\equiv\; M_{\rm Planck}^{-1}\, A_0\,, & h \;&\equiv\; M_{\rm Planck}^{-1}\, H\,, & \tau \;&\equiv\; M_{\rm Planck}\, t\,, \\ \lambda \;&\equiv\; M_{\rm Planck}^{-4}\, \Lambda\,, & r_{\rm m} \;&\equiv\; M_{\rm Planck}^{-4}\, \rho_{\rm m}\,, & p_{\rm m} \;&\equiv\; M_{\rm Planck}^{-4}\, P_{\rm m}\,. \end{aligned} \tag{5.15}$$

Thus, dot stands for a differentiation with respect to rescaled cosmic time τ throughout this section, rather than t. The rescaled energy density $r_v \equiv M_{\rm Planck}^{-4}\rho_{\rm A_0}$ and pressure $p_v \equiv M_{\rm Planck}^{-4}P_{\rm A_0}$ of the vector field are

$$\begin{aligned} r_v(\tau) &= -\zeta\left(\left(\dot{v}+3hv\right)^2 - 2v\left(\ddot{v}+3h\dot{v}_0+3\dot{h}v\right)\right) \\ &\quad +6\left(2\zeta_R-\zeta_S\right)\left(\dot{h}v^2-hv\dot{v}\right)+18\zeta_R h^2v^2\,, \end{aligned} \tag{5.16a}$$

$$\begin{aligned} p_v(\tau) &= +\zeta\Big(\left(\dot{v}+3hv\right)^2 + 2v\left(\ddot{v}+3h\dot{v}+3\dot{h}v\right)\Big) \\ &\quad +2\left(\zeta_R-\zeta_S\right)\left((2\dot{h}+3h^2)v^2+4hv\dot{v}\right)+2\left(2\zeta_R-\zeta_S\right)\left(\dot{v}^2+v\ddot{v}\right)\,. \end{aligned} \tag{5.16b}$$

Equation (5.13d) represents an evolution of the matter energy density with time, where we have taken $p_{\rm m} = w_{\rm m} r_{\rm m}$.

As mentioned in Section 2.4, $\nabla^\mu G_{\mu\nu} = 0$ due to the Bianchi identities. In the flat, homogeneous and isotropic universe it is equivalent to

$$\dot{G}_{00} + 3hG_{00} - hg^{ij}G_{ij} = 0\,, \tag{5.17}$$

where $G_{ij} \propto g_{ij}$ (see (2.5)). Hence, (5.13c) can be obtained by use of (5.13a) and (5.13b) as well as (5.13d) with an assumption $h \neq 0$. However, we do not leave it out from the equations (5.13), since it will be useful at certain places of our analysis below.

There are three separate subcases depending on a choice of the coefficients (5.5), namely: 1) $\zeta_S = \zeta_R = 0$; 2) $|\zeta_R|+|\zeta_S| \neq 0$ and $2\zeta_R-\zeta_S = 0$; 3) the rest of possibilities: $|\zeta_R|+|\zeta_S| \neq 0$ and $2\zeta_R-\zeta_S \neq 0$. Let us consider each of them in order.

First subcase: $\zeta_S = \zeta_R = 0$

Setting $\zeta_S = 0$ and $\zeta_R = 0$ in (5.13), these equations reduce to

$$\ddot{v} + 3h\dot{v} + 3\dot{h}v = 0\,, \tag{5.18a}$$

$$3h^2 - \lambda + \zeta\left(\dot{v}+3hv\right)^2 - r_{\rm m} = 0\,, \tag{5.18b}$$

$$\dot{r}_{\rm m} + 3h(1+w_{\rm m})r_{\rm m} = 0\,, \tag{5.18c}$$

where we have substituted $r_v(\tau)$ and omitted (5.13c) as needless.

The left-hand side of (5.18a) is a total derivative of $\dot{v} + 3hv$. Consequently, this must be a constant. Taking (5.18b) and (5.18c) into account, one concludes that either both $v(\tau)$ and $h(\tau)$ are constant or $v(\tau) \sim \tau$ and $h(\tau) \sim 1/\tau$. Specifically, we find two exact solutions:

$$v(\tau) = \bar{v}_0\,, \quad h(\tau) = \sqrt{\frac{\lambda}{3(1+3\zeta\bar{v}_0^2)}} \quad \text{and} \quad r_{\rm m}(\tau) = 0\,, \tag{5.19}$$

where $\bar{v}_0$ is a constant depending on initial conditions, and

$$v(\tau) = v_0\tau\,, \quad h(\tau) = \frac{2}{3(1+w_{\rm m})\tau} \quad \text{and} \quad r_{\rm m}(\tau) = \frac{4}{3(1+w_{\rm m})^2\tau^2}\,, \tag{5.20}$$

where v_0 have been defined as

$$v_0 \equiv \pm\frac{1+w_\mathrm{m}}{3+w_\mathrm{m}}\sqrt{\lambda/\zeta} \tag{5.21}$$

for $w_\mathrm{m} \neq -3$, otherwise there is no such solution.

Stability analysis In order to specify a solution of (5.18), one has to impose three initial conditions. For example, we can set the values of $v(\tau)$, $\dot{v}(\tau)$ and $r_\mathrm{m}(\tau)$ at some initial moment of time τ_in. However, (5.19) has only one arbitrary constant, but (5.20) is determined merely by the model parameters. It means that we have actually found particular exact solutions. It can be that a full solution of (5.18) behaves itself differently from the found ones at large time, so that the rest of independent solutions of (5.18) can manifest themselves in a way that (5.19) and (5.20) are spoilt at such time.

To make a stability analysis of both (5.19) and (5.20), we have to consider homogeneous perturbations around them, namely

$$v \to v+\delta v(\tau)\,, \quad h \to h+\delta h(\tau)\,, \quad r_\mathrm{m} \to r_\mathrm{m}+\delta r_\mathrm{m}(\tau)\,, \tag{5.22}$$

where $\delta v(\tau)$, $\delta h(\tau)$ and $\delta r_\mathrm{m}(\tau)$ are unknown functions. Then one needs to substitute (5.22) in (5.18) and linearize equations (5.18) with respect to the perturbations and look for their full solution. Depending on how $\delta v(\tau)$, $\delta h(\tau)$ and $\delta r_\mathrm{m}(\tau)$ behave themselves at large time τ, one can conclude whether the particular exact solutions found above are asymptotically stable or not.

First critical point Having linearized (5.18) with respect to $\delta v(\tau)$, $\delta h(\tau)$ and $\delta r_\mathrm{m}(\tau)$ around (5.19), one has

$$\delta\ddot{v} + 3h\delta\dot{v} + 3\bar{v}_0\delta\dot{h} = 0\,, \tag{5.23a}$$

$$6h\delta h - \delta r_\mathrm{m} + 6\zeta\, h\bar{v}_0\big(\delta\dot{v} + 3h\delta v + 3\bar{v}_0\delta h\big) = 0\,, \tag{5.23b}$$

$$\delta\dot{r}_\mathrm{m} + 3h\big(1+w_\mathrm{m}\big)\delta r_\mathrm{m} = 0\,. \tag{5.23c}$$

It is not hard to see that a solution of the last equation (5.23c) is

$$\delta r_\mathrm{m}(\tau) = 6\frac{h^2 w_\mathrm{m}}{\bar{v}_0}\,C_3 \exp\big(-3(1+w_\mathrm{m})h\tau\big)\,, \tag{5.24}$$

where C_3 is a constant of integration, while (5.23a) and (5.23b) give

$$\delta v(\tau) = C_1 + C_2 \exp\big(-3h\tau\big) + C_3 \exp\big(-3(1+w_\mathrm{m})h\tau\big)\,, \tag{5.25}$$

$$\delta h(\tau) = -\frac{3\zeta\, h\bar{v}_0}{1+3\zeta\bar{v}_0^2}\,C_1 + \frac{h w_\mathrm{m}}{\bar{v}_0}\,C_3 \exp\big(-3(1+w_\mathrm{m})h\tau\big)\,, \tag{5.26}$$

$C_{1,2}$ are arbitrary constants. Since the number of integration constants coincides with the number of initial conditions which must be imposed, we arrive at the conclusion that (5.24), (5.25) and (5.26) give a complete solution of (5.23).

For sufficiently large time $3(1+w_{\rm m})h\tau \gg 1$ subject to $w_{\rm m} > -1$, one has

$$\delta v(\tau) \approx C_1\,, \quad \delta h(\tau) \approx -\frac{3\zeta\, h\bar{v}_0}{1+3\zeta\bar{v}_0^2}\,C_1\,, \tag{5.27}$$

i.e. $\delta v(\tau)$ and $\delta h(\tau)$ are comparable with the background solutions $v(\tau)$ and $h(\tau)$, but $\delta r_{\rm m}(\tau)$ tends to zero. Moreover, it is straightforward to show that changing $\bar{v}_0$ into $\bar{v}_0 + \delta v(\tau)$ in $h(\tau)$ of (5.19), and then linearizing $h(\tau)$ with respect to $\delta v(\tau)$, we find that $\delta h(\tau)$ is exactly given by (5.27). It means, as already mentioned, the actual value of $\bar{v}_0$ as well as the value of $h(\tau)$ as a function of $\bar{v}_0$ are determined by the initial conditions. In other words, there exists a nonempty set of initial conditions, such that the solution of (5.18) approaches a curve in the three-dimensional phase space $(v, h, r_{\rm m})$, with an equation that is given by (5.19).

Second critical point Now let us take (5.20) as a background solution. The linearized equations (5.18) around (5.20) read

$$\delta\ddot{v} + \frac{2}{(1+w_{\rm m})\tau}\,\delta\dot{v} - \frac{2}{(1+w_{\rm m})\tau^2}\,\delta v + 3v_0\left(\tau\delta\dot{h} + \delta h\right) = 0\,, \tag{5.28a}$$

$$\frac{4}{(1+w_{\rm m})\tau}\,\delta h - \delta r_v - \delta r_{\rm m} = 0\,, \tag{5.28b}$$

$$\delta\dot{r}_{\rm m} + \frac{2}{\tau}\,\delta r_{\rm m} + \frac{4}{(1+w_{\rm m})\tau^2}\,\delta h = 0\,, \tag{5.28c}$$

where $\delta r_v(\tau)$ is the rescaled energy density of the vector field linearized with respect to the homogeneous perturbations,

$$\delta r_v(\tau) = -2\zeta v_0\frac{3+w_{\rm m}}{1+w_{\rm m}}\left(\delta\dot{v} + \frac{2}{(1+w_{\rm m})\tau}\,\delta v + 3v_0\tau\delta h\right)\,. \tag{5.29}$$

Although the equations (5.28) look, perhaps, complicated, they can be easily solved. Indeed, the first equation (5.28a) is a total time derivative of

$$\delta\dot{v} + \frac{2}{(1+w_{\rm m})\tau}\,\delta v + 3v_0\tau\delta h\,, \tag{5.30}$$

therefore this must be a constant, so that one obtains $\delta h(\tau)$ as a function of τ, $\delta v(\tau)$ and $\delta\dot{v}(\tau)$. Using this result, one gets from (5.28b) that $\delta r_{\rm m}(\tau)$ is also a function of τ, $\delta v(\tau)$ and $\delta\dot{v}(\tau)$. Thus, if we substitute $\delta h(\tau)$ and $\delta r_{\rm m}(\tau)$ in (5.28c), we derive a homogeneous linear differential equation of the second order in $\delta v(\tau)$ with time-variable coefficients. Having solved this equation, one has the following complete solution of the linearized system (5.28):

$$\delta v(\tau) = C_1 + C_2\tau^{-2/(1+w_{\rm m})} + C_3\left(\tau^2 + \frac{5+3w_{\rm m}}{\zeta\, v_0^2(3+w_{\rm m})^2}\right)\tau\,, \tag{5.31a}$$

$$\delta h(\tau) = -\frac{2C_1 + C_3(5+3w_{\rm m})\tau^3}{3v_0(1+w_{\rm m})\tau^2}\,, \tag{5.31b}$$

$$\delta r_{\rm m}(\tau) = -2\frac{4C_1 - C_3(5+3w_{\rm m})\tau^3}{3v_0(1+w_{\rm m})^2\tau^3}\,, \tag{5.31c}$$

where $C_{1,2,3}$ are constants of integration. Hence, for sufficiently large time

$$\delta v(\tau)/v(\tau) \sim \delta h(\tau)/h(\tau) \sim \delta r_{\rm m}(\tau)/r_{\rm m}(\tau) \sim C_3\tau^2\,, \tag{5.32}$$

i.e., generally speaking, the perturbations grow with time in comparison with (5.20). This means that (5.20) is an unstable solution of (5.18).

Note that C_1 can be interpreted as a shift of time. Indeed, it is straightforward to show that (5.20) with τ replaced by $\tau + C_1/v_0$ is also a solution of (5.18). So for large time $\tau \gg |C_1/v_0|$, one approximately has

$$\delta v(\tau) \approx C_1\,, \quad \delta h(\tau) \approx -\frac{2C_1}{3v_0(1+w_{\rm m})\tau^2}\,, \quad \delta r_{\rm m}(\tau) \approx -\frac{8C_1}{3v_0(1+w_{\rm m})^2\tau^3}\,. \tag{5.33}$$

This observation will be useful in the following.

Second subcase: $|\zeta_R| + |\zeta_S| \neq 0$ **and** $2\zeta_R - \zeta_S = 0$

Let us now treat the case where $\zeta_S = 2\zeta_R \neq 0$. Substituting this in (5.13), we have

$$\ddot v + 3h\dot v + 3\Big(\dot h + 2(\zeta_R/\zeta)h^2\Big)v = 0\,, \tag{5.34a}$$

$$3h^2 - \lambda - r_v - r_{\rm m} = 0\,, \tag{5.34b}$$

$$2\dot h + 3h^2 - \lambda + p_v + w_{\rm m} r_{\rm m} = 0\,, \tag{5.34c}$$

$$\dot r_{\rm m} + 3h(1+w_{\rm m})r_{\rm m} = 0\,, \tag{5.34d}$$

where the rescaled energy density (5.16a) and pressure (5.16b) of the vector field are

$$r_v(\tau) = -\zeta\big(\dot v + 3hv\big)^2 + 6\zeta_R(hv)^2\,, \tag{5.35a}$$

$$p_v(\tau) = +\zeta\big(\dot v + 3hv\big)^2 - 2\zeta_R\Big(\big(2\dot h + 9h^2\big)v^2 + 4hv\dot v\Big)\,. \tag{5.35b}$$

Here we do not omit (5.13c) from a reason to be made clear shortly.

One can easily see that (5.34) has two particular exact solutions corresponding to de Sitter spacetime

$$v(\tau) = 0\,, \quad h(\tau) = \sqrt{\lambda/3} \quad \text{and} \quad r_{\rm m}(\tau) = 0\,, \tag{5.36}$$

and Minkowski spacetime

$$v(\tau) = v_0\tau\,, \quad h(\tau) = 0 \quad \text{and} \quad r_{\rm m}(\tau) = 0\,, \tag{5.37}$$

where by definition $v_0 \equiv \pm\sqrt{\lambda/\zeta}$.

Stability analysis As in the previous subcase, we consider homogeneous perturbations around both (5.36) and (5.37) in order to analyze their stability.

de Sitter spacetime Considering the homogeneous perturbations of the variables around (5.36) and linearizing (5.34) with respect to them, we obtain

$$\delta\ddot{v} + 3h\delta\dot{v} + 6(\zeta_R/\zeta)h^2\delta v = 0\,, \tag{5.38a}$$

$$6h\delta h - \delta r_{\rm m} = 0\,, \tag{5.38b}$$

$$\delta\dot{r}_{\rm m} + 3h(1+w_{\rm m})\delta r_{\rm m} = 0\,. \tag{5.38c}$$

The complete solution of these equations is almost obvious, so we easily find

$$\delta v(\tau) = C_{1,2}\exp\left(-\frac{3h\tau}{2}\left(1 \pm \sqrt{1-8(\zeta_R/3\zeta)}\right)\right), \tag{5.39a}$$

$$\delta h(\tau) = C_3 \exp\left(-3(1+w_{\rm m})h\tau\right), \tag{5.39b}$$

$$\delta r_{\rm m}(\tau) = 6hC_3 \exp\left(-3(1+w_{\rm m})h\tau\right), \tag{5.39c}$$

where $C_{1,2,3}$ are arbitrary constants. We see that all of them approaches zero with growing time only if $w_{\rm m} > -1$ and

$$\zeta\zeta_R > 0 \tag{5.40}$$

is satisfied. If otherwise, the de Sitter spacetime solution is unstable.

Minkowski spacetime Analogously, linearizing (5.34) about the Minkowski spacetime solution (5.37) yields

$$\delta\ddot{v} + 3v_0\left(\tau\delta\dot{h} + \delta h\right) = 0\,, \tag{5.41a}$$

$$2\zeta v_0\left(\delta\dot{v} + 3v_0\tau\delta h\right) - \delta r_{\rm m} = 0\,, \tag{5.41b}$$

$$2\delta\dot{h} + 2\zeta v_0\left(\delta\dot{v} + 3v_0\tau\delta h\right) - 4\zeta_R v_0^2\tau^2\left(\delta\dot{h} + \frac{2}{\tau}\delta h\right) + w_{\rm m}\delta r_{\rm m} = 0\,, \tag{5.41c}$$

$$\delta\dot{r}_{\rm m} = 0\,. \tag{5.41d}$$

First, it directly follows from (5.41d) that $\delta r_{\rm m}(\tau) = \text{const}$. Second, (5.41a) is encoded in (5.41b) and (5.41d), therefore we can omit this. Then using (5.41b) and (5.41c), we obtain an equation in $\delta h(\tau)$ only. Specifically, one has

$$\delta\dot{h} - \frac{4\zeta_R v_0^2\tau}{1-2\zeta_R v_0^2\tau^2}\delta h = \frac{C_2}{1-2\zeta_R v_0^2\tau^2} \quad\Rightarrow\quad \delta h(\tau) = \frac{C_1 + C_2\tau}{1-2\zeta_R v_0^2\tau^2}\,, \tag{5.42}$$

where we have taken

$$\delta r_{\rm m}(\tau) = -\frac{2C_2}{1+w_{\rm m}}\,. \tag{5.43}$$

Substituting $\delta h(\tau)$ and $\delta r_{\rm m}(\tau)$ in (5.41b), we find

$$\delta v(\tau) = \frac{3\zeta(1+w_{\rm m})-2\zeta_R}{2\zeta\zeta_R v_0(1+w_{\rm m})}\,C_2\tau \qquad (5.44)$$

$$+\frac{3}{4v_0^2\zeta_R^{3/2}}\Big(C_1 v_0\zeta_R^{1/2}\ln\left|C_3(1-2\zeta_R v_0^2\tau^2)\right| - 2^{1/2}C_2\,\mathrm{Arctanh}\big(v_0\tau\sqrt{2\zeta_R}\big)\Big)\,.$$

Hence, Minkowski spacetime solution is unstable unless we remove the matter field from the model ($\psi = 0 \Rightarrow r_{\rm m} = p_{\rm m} = 0$). If so, then $C_2 = 0$ identically and, consequently, $\delta v(\tau)/v(\tau) \sim \ln(\tau)/\tau$ and $\delta h(\tau) \sim 1/\tau^2$ for large time $|\zeta_R|v_0^2\tau^2 \gg 1$.

Third subcase: $|\zeta_S| + |\zeta_R| \neq 0$ **and** $2\zeta_R - \zeta_S \neq 0$

One can directly show that the equations (5.13) have the following classes of particular exact solutions

$$v(\tau) = 0\,, \quad h(\tau) = \sqrt{\lambda/3} \quad \text{and} \quad r_{\rm m}(\tau) = 0 \qquad (5.45)$$

corresponding to de Sitter spacetime, and

$$v(\tau) = v_0\tau\,, \quad h(\tau) = \frac{2}{3(1+w_{\rm m})\tau} \quad \text{and} \quad r_{\rm m}(\tau) = \frac{4}{3(1+w_{\rm m})^2\tau^2}\,, \qquad (5.46)$$

where v_0 have been defined as[(1)]

$$v_0 = \pm\left(\frac{\lambda(4\zeta_R-\zeta_S)^2}{2(5\zeta_R-2\zeta_S)\big(3(2\zeta_R-\zeta_S)^2+2\zeta(5\zeta_R-2\zeta_S)\big)}\right)^{1/2} \qquad (5.47)$$

and the constant parameter of the equation of state is given by

$$w_{\rm m} = \frac{1}{3}+\frac{2\zeta_S}{3(2\zeta_R-\zeta_S)}\,, \qquad (5.48)$$

i.e. $w_{\rm m}$ is fixed by the ratio ζ_S/ζ_R.

Note that Dolgov's model [5] corresponds to $\zeta_S = \zeta_F \neq 0$, $\zeta_Q = \zeta_R = 0$ and, hence, $w_{\rm m} = -1/3$ as in a universe filled with a string gas. The model considered in [98] corresponds to $\zeta_S = \zeta_F = 0$, $\zeta_Q \neq 0$ and $\zeta_R \neq 0$, such as $w_{\rm m} = 1/3$ as for radiation.

Stability analysis

---

(1) We tacitly exclude values of the ζ-coefficients throughout our work which imply infinite or imaginary values of the dynamical variables. In particular, the dominator in (5.47) is nonzero and v_0 is assumed to be real.

First critical point The differential equations (5.13) linearized with respect to the homogeneous perturbations $\delta v(\tau)$, $\delta h(\tau)$ and $\delta r_{\rm m}(\tau)$ around (5.45) are

$$\delta\ddot{v} + 3h\delta\dot{v} - 9(\beta/\alpha)h^2\delta v = 0\,, \tag{5.49a}$$

$$6h\delta h - \delta r_{\rm m} = 0\,, \tag{5.49b}$$

$$\delta\dot{r}_{\rm m} + 3(1+w_{\rm m})h\delta r_{\rm m} = 0\,, \tag{5.49c}$$

where we have omitted (5.13c) as needless here and defined new parameters as follows

$$\alpha \equiv -3\frac{2\zeta_R - \zeta_S}{4\zeta_R - \zeta_S} \neq 0\,, \quad \beta \equiv \zeta^{-1}(2\zeta_R - \zeta_S) \neq 0\,, \tag{5.50}$$

in order to make the further analysis more transparent.[(2)] It is straightforward to find a complete solution of (5.49). Indeed, equation (5.49a) is satisfied by

$$\delta v(\tau) = C_{1,2}\exp\left(-\frac{3h\tau}{2}\left(1 \pm \sqrt{1+4(\beta/\alpha)}\right)\right), \tag{5.51}$$

while the other two (5.49b) and (5.49c) are solved by

$$\delta h(\tau) = C_3 \exp\left(-3(1+w_{\rm m})h\tau\right), \tag{5.52}$$

$$\delta r_{\rm m}(\tau) = 6hC_3 \exp\left(-3(1+w_{\rm m})h\tau\right). \tag{5.53}$$

We see that only if $w_{\rm m} > -1$ and $\beta/\alpha < 0$ or, equivalently,

$$\zeta(4\zeta_R - \zeta_S) > 0\,, \tag{5.54}$$

then the perturbations $\delta v(\tau)$, $\delta h(\tau)$ and $\delta r_{\rm m}(\tau)$ tend to zero when time approaches infinity and this does not depend on the constants $C_{1,2,3}$, so that we conclude that de Sitter spacetime solution is asymptotically stable only if (5.54) holds and $w_{\rm m} > -1$.

Now let us assume $\beta/\alpha > 0$ and $|\beta/\alpha| \ll 1$. Since

$$\sqrt{1-4\,|\beta/\alpha|} \approx 1 + 2\,|\beta/\alpha|\,, \tag{5.55}$$

we have for sufficiently large time

$$\delta v(\tau) \approx C_2 \exp\left(3\,|\beta/\alpha|\,h\tau\right). \tag{5.56}$$

In other words, there exists a phase of de Sitter spacetime which lasts till $\tau_{\rm dS}$ that is approximately equal to

$$\tau_{\rm dS} \sim \frac{1}{h}|\alpha/\beta| = \frac{1}{h}\left|\frac{3\zeta}{4\zeta_R - \zeta_S}\right|. \tag{5.57}$$

After this moment of time, $\delta v(\tau)$ changes considerably as well as $\delta h(\tau)$ and $\delta r_{\rm m}(\tau)$, so that the linearized equations (5.49) are not reliable anymore. The actual moment of time, at which the homogeneous

[(2)]Note, both α and β are nonzero, since $2\zeta_R \neq \zeta_S$ in the present subcase.

perturbations of the dynamical variables change dramatically, however, strongly depends on the initial conditions, so that an actual duration of this phase can be different from $\tau_{\rm dS}$, but if one keeps the initial conditions unchanged and increases, for example, ζ, de Sitter phase becomes longer in accordance with (5.57) [99].

Second critical point The differential equations (5.13) linearized with respect to the perturbations $\delta v(\tau)$, $\delta h(\tau)$ and $\delta r_{\rm m}(\tau)$ around (5.46) read

$$\delta\ddot{v} - \frac{\alpha}{\tau}\,\delta\dot{v} + \frac{\alpha}{\tau^2}\,\delta v + 3v_0\left((\beta+1)\tau\delta\dot{h} + (2\beta+1)\delta h\right) = 0\,, \tag{5.58a}$$

$$6h\delta h - \delta r_v - \delta r_{\rm m} = 0\,, \tag{5.58b}$$

$$2\delta\dot{h} + 6h\delta h + \delta p_v + w_{\rm m}\delta r_{\rm m} = 0\,, \tag{5.58c}$$

$$\delta\dot{r}_{\rm m} + \frac{2}{\tau}\,\delta r_{\rm m} - \frac{2\alpha}{\tau^2}\,\delta h = 0\,, \tag{5.58d}$$

where the rescaled and linearized energy density and pressure of the vector field are

$$\begin{aligned}(\zeta\, v_0)^{-1}\delta r_v(\tau) &= 2\tau\delta\ddot{v} + 2(\alpha\beta-1)\delta\dot{v} - \frac{2\alpha}{\tau}(\alpha\beta+\alpha-2)\delta v \\ &\quad +6v_0\left((\beta+1)\tau^2\delta\dot{h} + (\alpha\beta+\alpha+2\beta)\tau\delta h\right)\,,\end{aligned} \tag{5.59}$$

$$\begin{aligned}(\zeta\, v_0)^{-1}\delta p_v(\tau) &= 2(\beta+1)\delta\ddot{v} - 2(2\alpha\beta+2\alpha-1)\delta\dot{v} + \frac{2\alpha}{\tau}(\alpha\beta+\alpha+\beta)\delta v \\ &\quad +\frac{6v_0}{\alpha}(\alpha\beta+\alpha+\beta)\left(\tau^2\delta\dot{h} - (\alpha-2)\tau\delta h\right)\,.\end{aligned} \tag{5.60}$$

From the system of equations (5.58), one can extract a differential equation in $\delta v(\tau)$ only, such that the rest of unknown functions, i.e. $\delta h(\tau)$ and $\delta r_{\rm m}(\tau)$, are determined by $\delta v(\tau)$ and its first and second derivatives. Specifically, from (5.58b) we obtain $\delta r_{\rm m}(\tau)$ as a function of τ, $\delta h(\tau)$, $\delta\dot{h}(\tau)$ and $\delta v(\tau)$, $\delta\dot{v}(\tau)$, $\delta\ddot{v}(\tau)$. Now substituting this in (5.58c), one gets $\delta\dot{h}(\tau)$ as a function of τ, $\delta h(\tau)$, $\delta v(\tau)$ and its first and second derivatives. Thus, we find $\delta h(\tau)$ as a function of τ and $\delta v(\tau)$, $\delta\dot{v}(\tau)$, $\delta\ddot{v}$ and $\delta\dddot{v}(\tau)$ from (5.58d). And, finally, with the help of (5.58a), we get

$$\delta\dddot{v} + a_1(\tau)\,\delta\ddot{v} + a_2(\tau)\,\delta\dot{v} = 0\,, \tag{5.61}$$

where $a_1(\tau)$ and $a_2(\tau)$ have been defined as

$$a_1(\tau) \equiv -\frac{\beta(\alpha-2) + 2(\alpha+3\beta)(v_0\eta\tau)^2 - (\alpha-4)(2+\beta)(v_0\eta\tau)^4}{\tau\left(\beta + 2(v_0\eta\tau)^2 - (2+\beta)(v_0\eta\tau)^4\right)}\,, \tag{5.62a}$$

$$a_2(\tau) \equiv \frac{2(\alpha-1)(v_0\eta)^2\left(2+3\beta+(2+\beta)(v_0\eta\tau)^2\right)}{\left(1-(v_0\eta\tau)^2\right)\left(\beta+(2+\beta)(v_0\eta\tau)^2\right)}\,, \tag{5.62b}$$

where

$$\eta \equiv \left(3\alpha^{-1}\beta\zeta(\alpha+\alpha\beta-1)\right)^{1/2}\,. \tag{5.63}$$

One can directly verify that both C_1 and $C_2\tau^\alpha$ satisfy the equation (5.61). The third independent solution can be found by using the well-known Liouville-Ostrogradsky formula. Having applied that, we obtain the following exact solution of the linearized system (5.58):

$$\begin{aligned}\delta v(\tau) &= C_1 + C_2\tau^\alpha + C_3\,\mathrm{Arctanh}\big(v_0\eta\tau\big) \\ &\quad - \frac{2\alpha\eta^2 C_3}{3\zeta\beta(\alpha-1)}(v_0\eta\tau)\,{}_2F_1\left(1,\ \frac{1-\alpha}{2};\ \frac{3-\alpha}{2};\ (v_0\eta\tau)^2\right),\end{aligned} \tag{5.64}$$

where ${}_2F_1(a,b;c;z)$ is the hypergeometric series of variable z with parameters a, b and c [100], where $\alpha \neq 3$, and

$$\delta h(\tau) = \frac{\alpha}{3v_0\tau^2}\left(C_1 + C_3\left(\ln\left(\frac{1+v_0\eta\tau}{1-v_0\eta\tau}\right) - \frac{2v_0\eta\tau}{1-(v_0\eta\tau)^2}\right)\right), \tag{5.65}$$

$$\delta r_{\mathrm{m}}(\tau) = -\frac{2\alpha^2}{3v_0\tau^3}\left(C_1 + C_3\left(\ln\left(\frac{1+v_0\eta\tau}{1-v_0\eta\tau}\right) - 2v_0\eta\tau\right)\right). \tag{5.66}$$

We are interested in the behavior of the perturbations when time is large $|v_0\eta\tau| \gg 1$. Therefore, expanding them in a series, one obtains

$$\delta v(\tau) = \widetilde{C}_1 + \widetilde{C}_2\tau^\alpha + \frac{\alpha(\beta+2)}{v_0(\alpha+1)\tau}\widetilde{C}_3 + \mathrm{O}\left(\frac{1}{\tau^3}\right), \tag{5.67a}$$

$$\delta h(\tau) = \frac{\alpha}{3v_0\tau^2}\left(\widetilde{C}_1 + \frac{2\widetilde{C}_3}{3v_0\tau^2} + \mathrm{O}\left(\frac{1}{\tau^3}\right)\right), \tag{5.67b}$$

$$\delta r_{\mathrm{m}}(\tau) = \frac{2(\alpha\eta)^2}{3\tau^2}\left(\widetilde{C}_3 - \frac{\widetilde{C}_1}{v_0\eta^2\tau} + \mathrm{O}\left(\frac{1}{\tau^2}\right)\right), \tag{5.67c}$$

where we have made redefinitions of the integration constants in such a way to have a real solution for real integration constants.

First, we see that $\widetilde{C}_1$ can be interpreted as a shift of time (see above). Second, for $\delta v(\tau)/v(\tau)$ to go to zero with growing time, the term $\widetilde{C}_2\tau^\alpha$ in $\delta v(\tau)$ must be small in comparison with $\widetilde{C}_1$. In other words, α must be negative.(3) This condition in terms of ζ_S and ζ_R reads

$$\zeta_S/\zeta_R \in D, \quad \text{where} \quad D = \big(-\infty, 2\big) \cup \big(4, +\infty\big). \tag{5.68}$$

Thirdly, the ratio $\delta h(\tau)/h(\tau)$ approaches zero as $1/\tau^3$. Fourthly, $\delta r_{\mathrm{m}}(\tau)/r_{\mathrm{m}}(\tau) \to$ const. Since the equations are linear with respect to $r_{\mathrm{m}}(\tau)$, such behavior for $\delta r_{\mathrm{m}}(\tau)$ does not invalidate the solution. It just tells us that the final value of $r_{\mathrm{m}}(\tau)$ depends on the initial conditions, while final values of $v(\tau)$ and $h(\tau)$ are entirely determined by the parameters of the model.

(3)In principle, if $0 < \alpha < 1$, then $\delta v(\tau)/v(\tau)$ still approaches zero, when time tends to infinity. However, the perturbation of the vector energy density $\delta r_v(\tau)$ decreases merely as $\tau^{2(\alpha-1)}$, i.e. it becomes dominant at large time in comparison with $h^2 \sim \tau^{-2}$ and $r_{\mathrm{m}} \sim \tau^{-2}$ in the "time-time" Einstein equation and eventually destroys the background solution under consideration.

It means there exists a domain of initial conditions in the phase space of the variables, such that asymptotic behavior of both $v(\tau)$ and $h(\tau)$ are independent of which point from this domain we take.

The physically relevant values of the constant parameter of state $w_{\rm m}$ belongs to the half-interval $(-1,\, 1]$. In terms of the ratio ζ_S/ζ_R according to (5.48), this becomes

$$\zeta_S/\zeta_R \;\in\; \widetilde{D}, \quad \text{where} \quad \widetilde{D} \;=\; \big(-\infty,\, 1\big] \cup \big(4,\, +\infty\big)\,. \tag{5.69}$$

Clearly, $\widetilde{D} \in D$, so that for any physically relevant matter field, the cosmological constant is compensated by the time component of the vector field.

5.2.2 Case: $A_\mu = (0, A_i)$

Let us look for a particular exact solution of $A_i(\tau) = M_{\rm Planck}\, a_i(\tau)$ in the following form

$$a_i(\tau) \;=\; a(\tau)\chi(\tau)\xi_i, \tag{5.70}$$

where, we recall, $a(\tau)$ is the scale factor, ξ_i is a unit constant three-dimensional vector, and $\chi(\tau)$ is a dimensionless function describing an evolution of the spatial component of the vector field.

The vector field equation of $A_i(\tau)$ rewritten via the dimensionaless function $\chi(\tau)$ reads

$$\ddot{\chi} + 3h\dot{\chi} + \Big(\dot{h} + 2h^2 + 2\bar{\zeta}^{-1}\Big[\big(6\zeta_R - \zeta_S\big)\dot{h} + 3\big(4\zeta_R - \zeta_S\big)h^2\Big]\Big)\chi \;=\; 0\,, \tag{5.71}$$

where by definition

$$\bar{\zeta} \;\equiv\; \zeta_F + \zeta_S\,. \tag{5.72}$$

Note, we do not assume that $\bar{\zeta}$ cannot be zero.

The Einstein equations in components and the equation of the evolution of the matter energy density are

$$3h^2 - \lambda - r_\chi - r_{\rm m} \;=\; 0\,, \tag{5.73a}$$

$$2\dot{h} + 3h^2 - \lambda + p_\chi + w_{\rm m} r_{\rm m} \;=\; 0\,, \tag{5.73b}$$

$$\dot{r}_{\rm m} + 3h(1 + w_{\rm m}) r_{\rm m} \;=\; 0\,, \tag{5.73c}$$

where $r_\chi(\tau)$ and $p_\chi(\tau)$ are parts of the rescaled energy density and pressure of A_μ associated with its spatial components $A_i(\tau)$ expressed via $\chi(\tau)$:

$$r_\chi \;=\; \frac{\bar{\zeta}}{2}\big(\dot{\chi} + h\chi\big)^2 + 2\big(6\zeta_R - \zeta_S\big)h\chi\dot{\chi} + 6\zeta_R h^2\chi^2\,, \tag{5.74a}$$

$$p_\chi \;=\; \frac{\bar{\zeta}}{2}\big(\dot{\chi} + h\chi\big)^2 - 2\big(4\zeta_R + \zeta_S\big)h\chi\dot{\chi} - 2\zeta_R\Big(\big(2\dot{h} + 3h^2\big)\chi^2 + 2\dot{\chi}^2 + 2\chi\ddot{\chi}\Big)\,. \tag{5.74b}$$

As mentioned, when $A_i \neq 0$, the non-diagonal elements of $T^{\mathrm{A}}_{\mu\nu}$ do not vanish identically. This constrains the allowed time-dependence of A_i, namely we have an extra equation resulting from $T^{\mathrm{A}}_{ij} = 0$ for $i \neq j$, which in terms of $\chi(\tau)$ is given by

$$
\begin{aligned}
0 &= \zeta_S(\ddot{\chi} + h\dot{\chi})\chi \\
&\quad + \left(\zeta_S - \frac{\bar{\zeta}}{2}\right)(\dot{\chi} + h\chi)^2 + \left((6\zeta_R - \zeta_S)\dot{h} + 4(3\zeta_R - \zeta_S)h^2\right)\chi^2
\end{aligned}
\tag{5.75}
$$

(see Appendix C for more details).

If $\chi(\tau) = 0$, then (5.71) is clearly satisfied. In addition, $r_\chi(\tau)$ and $p_\chi(\tau)$ as well as (5.75) are exactly zero. Hence, we arrive at a simple conclusion that

$$
\chi(\tau) = 0, \quad h(\tau) = \sqrt{\lambda/3} \quad \text{and} \quad r_{\mathrm{m}}(\tau) = 0 \tag{5.76}
$$

satisfy simultaneously all equations we have. This particular exact solution corresponds to de Sitter universe.

In order to cancel the cosmological constant λ in (5.73a) by the vector field only, $\chi(\tau)$ must linearly increase with time. Indeed, de Sitter spacetime is excluded, so that assuming $h(\tau) \sim 1/\tau$,(4) one can see that $r_\chi(\tau)$ is a constant only if $\chi(\tau) \sim \tau$. It is straightforward to verify that if

$$
\bar{\zeta} = \zeta_S \frac{4\zeta_R - \zeta_S}{3\zeta_R - \zeta_S} \tag{5.77}
$$

holds, then

$$
\chi(\tau) = \chi_0 \tau, \quad h(\tau) = \frac{2}{3(1 + w_{\mathrm{m}})\tau} \quad \text{and} \quad r_{\mathrm{m}}(\tau) = \frac{4}{3(1 + w_{\mathrm{m}})^2 \tau^2} \tag{5.78}
$$

is one more exact particular solution of (5.71), (5.73a)–(5.73c) as well as (5.75), where we have defined χ_0 as

$$
\chi_0 = \pm \left(\frac{\lambda (4\zeta_R - \zeta_S)^2}{8\zeta_R(\zeta_S - 3\zeta_R)(5\zeta_R - 2\zeta_S)} \right)^{1/2} \tag{5.79}
$$

and w_{m} is determined by ζ_S and ζ_R in the same manner as in (5.48).

Stability analysis

First critical point The small perturbations $\delta\chi(\tau)$, $\delta h(\tau)$ and $\delta r_{\mathrm{m}}(\tau)$ around (5.76) evolve according to

$$
\delta\ddot{\chi} + 3h\delta\dot{\chi} + 2\left(1 + 3\bar{\zeta}^{-1}(4\zeta_R - \zeta_S)\right)h^2\delta\chi = 0\,, \tag{5.80a}
$$

$$
6h\delta h - \delta r_{\mathrm{m}} = 0\,, \tag{5.80b}
$$

$$
\delta\dot{r}_{\mathrm{m}} + 3(1 + w_{\mathrm{m}})h\delta r_{\mathrm{m}} = 0\,. \tag{5.80c}
$$

(4)This comes from a dimensional consideration of $H(t)$.

We have omitted here the linearized versions of (5.73b) and (5.75), since they are automatically satisfied.

It is easy to show that

$$\delta\chi(\tau) = C_{1,2}\exp\left(-\frac{3h\tau}{2}\left(1\pm\frac{1}{3}\sqrt{1-24\bar{\zeta}^{-1}(4\zeta_R-\zeta_S)}\right)\right), \tag{5.81a}$$

$$\delta h(\tau) = C_3\exp\left(-3(1+w_m)h\tau\right), \tag{5.81b}$$

$$\delta r_{\rm m}(\tau) = 6hC_3\exp\left(-3(1+w_m)h\tau\right) \tag{5.81c}$$

give a complete solution of (5.80). Consequently, de Sitter spacetime solution (5.76) is asymptotically stable only if $w_{\rm m} > -1$ and

$$\bar{\zeta}\left(\bar{\zeta}+12\zeta_R-3\zeta_S\right) > 0 \tag{5.82}$$

are satisfied.

Second critical point Considering $\delta\chi(\tau)$, $\delta h(\tau)$ and $\delta r_{\rm m}(\tau)$ around (5.78), we find that they are governed by

$$\delta\ddot{\chi}-\frac{\alpha}{\tau}\delta\dot{\chi}+\frac{\alpha}{\tau^2}\delta\chi+\chi_0\frac{27+3\alpha-\alpha^2}{3(2\alpha+3)}\left(\tau\delta\dot{h}+\frac{27-12\alpha-2\alpha^2}{27+3\alpha-\alpha^2}\delta h\right) = 0, \tag{5.83a}$$

$$\frac{2\alpha}{\tau}\delta h+\delta r_\chi+\delta r_M = 0, \tag{5.83b}$$

$$2\delta\dot{h}-\frac{2\alpha}{\tau}\delta h+\delta p_\chi+w_{\rm m}\delta r_{\rm m} = 0, \tag{5.83c}$$

$$\delta\dot{r}_{\rm m}+\frac{2}{\tau}\delta r_{\rm m}-\frac{2\alpha}{\tau^2}\delta h = 0, \tag{5.83d}$$

where the linearized quantities $r_\chi(\tau)$ and $p_\chi(\tau)$ are given by

$$\gamma^{-1}\delta r_\chi(\tau) = \frac{4}{3}(\alpha-3)\left(\delta\dot{\chi}-\frac{\alpha}{\tau}\delta\chi+3\chi_0\tau\delta h\right), \tag{5.84}$$

$$\gamma^{-1}\delta p_\chi(\tau) = 4\tau\delta\ddot{\chi}-\frac{4}{3}(4\alpha-3)\delta\dot{\chi}+\frac{4\alpha^2}{3\tau}\delta\chi+4\chi_0\tau\left(\tau\delta\dot{h}-(\alpha-2)\delta h\right). \tag{5.85}$$

Here by definition

$$\gamma \equiv -\chi_0\,\zeta_R\,. \tag{5.86}$$

By using (5.83b), we can find $\delta r_{\rm m}(\tau)$ as a function of τ, $\delta h(\tau)$, $\delta\chi(\tau)$ and $\delta\dot{\chi}(\tau)$. Then if we substitute this in (5.83c), we get $\delta\dot{h}(\tau)$ as a function of τ, $\delta h(\tau)$, $\delta\chi(\tau)$ and its first and second derivatives. After that, we can find $\delta h(\tau)$ from (5.83a) as a function of $\delta\chi(\tau)$ and its derivatives only, so that $\delta h(\tau)$ and $\delta r_{\rm m}(\tau)$ can be entirely expressed via $\delta\chi(\tau)$, $\delta\dot{\chi}(\tau)$ and $\delta\ddot{\chi}(\tau)$. Thus, if we now substitute $\delta h(\tau)$ and $\delta r_{\rm m}(\tau)$ in (5.83d), we obtain

$$\delta\dddot{\chi}+b_1(\tau)\,\delta\ddot{\chi}+b_2(\tau)\,\delta\dot{\chi} = 0\,. \tag{5.87}$$

Here $b_1(\tau)$ and $b_2(\tau)$ are known functions of time τ. They are given by huge expressions, therefore we do not write them down. Since we are interested in the asymptotic behavior of the perturbations, we expand $b_1(\tau)$ and $b_2(\tau)$ in a series, where τ is large. So we approximately obtain[(5)]

$$b_1(\tau) = \frac{4-\alpha}{\tau} + \mathrm{O}\left(\frac{1}{\tau^3}\right), \tag{5.88a}$$

$$b_2(\tau) = \frac{2(1-\alpha)}{\tau^2} + \mathrm{O}\left(\frac{1}{\tau^4}\right). \tag{5.88b}$$

Now substituting (5.88a) and (5.88b) in (5.87), one eventually finds

$$\delta\chi(\tau) = C_1 + C_2\tau^\alpha + \frac{C_3}{\tau} + \mathrm{O}\left(\frac{1}{\tau^3}\right), \tag{5.89a}$$

$$\delta h(\tau) = \frac{\alpha}{3\chi_0\tau^2}\left(C_1 + 2C_3\frac{(\alpha+1)(\alpha+9)(2\alpha+3)}{(54+24\alpha+\alpha^2)\tau} + \mathrm{O}\left(\frac{1}{\tau^3}\right)\right), \tag{5.89b}$$

$$\delta r_{\mathrm{m}}(\tau) = -4\gamma\alpha\, C_3\frac{(\alpha+1)(\alpha+6)(\alpha-3)}{(54+24\alpha+\alpha^2)\tau^2} - \frac{2\alpha^2C_1}{3\chi_0\tau^3} + \mathrm{O}\left(\frac{1}{\tau^4}\right). \tag{5.89c}$$

This is a complete solution of the system (5.83) in the limit of large time. However, we have not taken into account (5.75). This is incorrect, because this additional equation makes a constraint on the perturbations $\delta\chi(\tau)$, $\delta h(\tau)$ and $\delta r_{\mathrm{m}}(\tau)$, so that it could be that there does not exist a solution of all equations at all. However, this is not the case, since linearizing (5.75) yields

$$\delta\ddot{\chi} - \frac{\alpha}{\tau}\delta\dot{\chi} + \frac{\alpha}{\tau^2}\delta\chi + \chi_0\frac{6+\alpha}{3+2\alpha}\left(\tau\delta\dot{h} + \frac{3-4\alpha}{6+\alpha}\delta h\right) = 0, \tag{5.90}$$

which is satisfied by (5.89a), (5.89b) and (5.89c) if merely only one of the integration constants is zero, namely C_3. Therefore, omitting C_3 in (5.89), we obtain now a full exact[(6)] solution of (5.83) with (5.90).

As in Subsection 5.2.1, (5.78) is asymptotically stable only if (5.68) holds.

5.2.3 Case: $A_\mu = (A_0, A_i)$

De Sitter spacetime

Stability analysis made in Subsections 5.2.1 and 5.2.2 show that de Sitter spacetime solution

$$v(\tau) = \chi(\tau) = 0, \quad h(\tau) = \sqrt{\lambda/3}, \quad r_{\mathrm{m}}(\tau) = 0 \tag{5.91}$$

is asymptotically stable only if both conditions (5.54) and (5.82) are satisfied, and $w_{\mathrm{m}} > -1$.

[(5)]One could obtain an exact expression for $\delta\chi$ from (5.87) by taking into account that C_1 and $C_2\tau^\alpha$ are two of three independent exact solutions and then applying the Liouville-Ostrogradsky formula. Direct calculations show $\delta\chi$ is expressed in terms of the hypergeometric function of two variables. This causes unnecessary difficulties, therefore we consider the limit of large time before solving (5.87) for simplicity.

[(6)]See previous footnote.

Spacetime with compensated cosmological constant by vector field

We have seen that one is able to dynamically cancel the rescaled cosmological constant λ by the vector field in both previous cases under certain conditions uncovered above. It turned out that $v(\tau)$ or $\chi(\tau)$ must linearly grow with time, while $h(\tau) \sim 1/\tau$ and $r_{\rm m}(\tau) \sim 1/\tau^2$. We have ignored the spatial components of the vector field in the first case and its time component in the second case. This is inconsistent, therefore, let us consider them simultaneously when $|\zeta_S| + |\zeta_R| \neq 0$ and $2\zeta_R - \zeta_S \neq 0$.[(7)]

At first, let us treat (5.46) with $\chi(\tau) = 0$. It follows from (5.71), that $\delta\chi(\tau)$ evolves according to

$$\delta\ddot{\chi} - \frac{\alpha}{\tau}\,\delta\dot{\chi} + \frac{\alpha(2\alpha+3)}{9\tau^2}\left(1 + \frac{24\zeta_R}{\bar{\zeta}(\alpha+3)}\right)\delta\chi \;=\; 0 \tag{5.92}$$

in the linear approximation. Looking for a solution in a form $\delta\chi(\tau) \sim \tau^z$, we obtain

$$z^{\pm} \;=\; \frac{\alpha+1}{2} \pm \frac{\alpha+3}{6}\left(1 - 96\frac{\alpha\zeta_R}{\bar{\zeta}}\frac{2\alpha+3}{(\alpha+3)^3}\right)^{1/2}. \tag{5.93}$$

If Re$(z^{\pm}) < 0$, then $v(\tau) \sim \tau$, $\chi(\tau) = 0$, $h(\tau) \sim 1/\tau$ and $r_{\rm m}(\tau) \sim 1/\tau^2$ is an asymptotically stable solution. As an example, let us put $\zeta_S = \zeta_F = 1$ and $\zeta_Q = \zeta_R = 0$. This set of the coefficients corresponds to Dolgov's model. Substituting them in (5.93), one calculates $z^{\pm} = -1 \pm \sqrt{2}$. Consequently, the Dolgov vector model is unstable. This is actually the third flaw of the model discussed by Dolgov himself in [5]. If we set $\zeta_S = -2\zeta_R$ (this corresponds to $w_{\rm m} = 0$) and $\zeta_F/\zeta_R \in (-\infty,\, -22) \cup (2,\, +\infty)$, then Re$(z^{\pm}) < 0$.

As concerns (5.78) with $v(\tau) = 0$, it is quite clear that its homogeneous perturbation $\delta v(\tau)$ is a solution of the following equation

$$\delta\ddot{v} - \frac{\alpha}{\tau}\,\delta\dot{v} + \frac{\alpha}{\tau^2}\,\delta v \;=\; 0 \quad\Rightarrow\quad \delta v(\tau) \;=\; C_1\tau + C_2\tau^{\alpha}\,, \tag{5.94}$$

where $C_{1,2,}$ are constants of integration. Since $\delta v(\tau) \sim \tau$ for large time, (5.78) with $v(\tau) = 0$ is unstable solution.

Now let us suppose $v(\tau) = v_0\tau$, $\chi(\tau) = \chi_0\tau$ and $h(\tau) \sim 1/\tau$, i.e. we assume that (5.77) is satisfied. Before we start, let us figure out how many initial conditions we have to impose in order to specify a solution. From the vector field equations (5.13a) and (5.71), we have $\ddot{v} = f_v(v, \dot{v}, h, \dot{h})$ and $\ddot{\chi} = f_\chi(\chi, \dot{\chi}, h, \dot{h})$ subject to $\bar{\zeta} \neq 0$. Whence we can obtain $\dot{h} = f_{\dot{h}}(\chi, \dot{\chi}, h)$ from (5.75). So $r_{\rm m} = f_{r_{\rm m}}(v, \dot{v}, \chi, \dot{\chi}, h)$ can be found from the "time-time" Einstein equation and, consequently, $h = f_h(v, \dot{v}, \chi, \dot{\chi})$ from the "space-space" Einstein equation. Thus, we must impose 4 initial conditions, say, initial values of $v(\tau)$, $\dot{v}(\tau)$, $\chi(\tau)$ and $\dot{\chi}(\tau)$ at certain initial moment of time $\tau_{\rm in}$.

Subtracting (5.83a) from (5.90), one obtains

$$\delta\dot{h} + \frac{2}{\tau}\,\delta h \;=\; 0 \quad\Rightarrow\quad \delta h(\tau) \;=\; \frac{\alpha\, C_1}{3\tau^2}\,. \tag{5.95}$$

---

(7) Note, if $\zeta_S = \zeta_R = 0$, then (5.20) with $\chi(\tau) \propto \tau^{-2/3(1+w_{\rm m})}$ is a particular exact solution. If $\zeta_S = 2\zeta_R$ and $\zeta_R \neq 0$, then (5.37) with $\chi(\tau) = \text{const}$ is a particular solution as well. We exclude both these cases in what follows, because these solutions are not asymptotically stable.

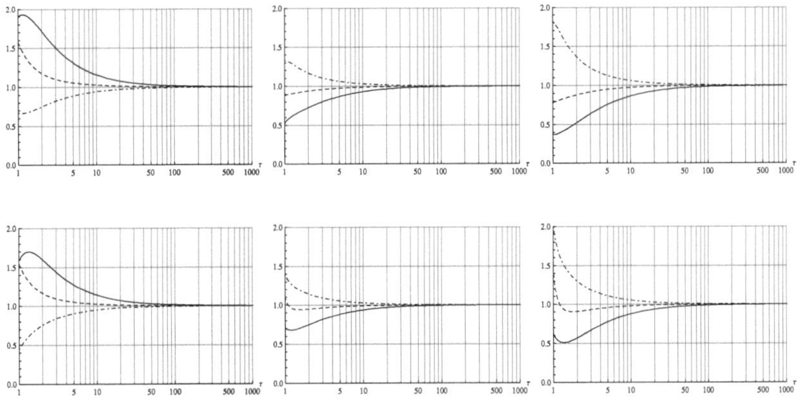

Figure 5.1: Numerical solutions of vector and Einstein field equations. The first column shows $-\lambda^{-1}(r_v(\tau) + r_\chi(\tau))$ as a function of rescaled time τ for three different sets of initial conditions. The second and third columns show $\frac{3}{2}\tau h(\tau)$ and $\frac{3}{4}\tau^2 r_{\rm m}(\tau)$, respectively. The top and bottom lines differ from each other by the initial conditions and the values of the model parameters, namely the top line corresponds to $(\lambda, \zeta, \bar{\zeta}, \zeta_S, \zeta_R) = (-1, -1, -12/5, -2, +1)$, while the bottom line – $(\lambda, \zeta, \bar{\zeta}, \zeta_S, \zeta_R) = (+1, -1, +12/5, +2, -1)$. For both lines $w_{\rm m}$ equals 0. Our numerical calculations are in complete agreement with (5.98) and (5.97), i.e. the value of $\tau^2 r_{\rm m}(\tau)$ at $\tau \to \infty$ does not depend on the initial conditions.

Substituting this in (5.58a) and (5.83a), one has

$$\delta v(\tau) = v_0 C_1 + C_2 \tau^\alpha + C_4 \beta(\alpha^2 - 9)\chi_0 \tau\,, \tag{5.96a}$$

$$\delta \chi(\tau) = \chi_0 C_1 + C_3 \tau^\alpha - 3C_4 \alpha(\alpha\beta + \alpha - 1) v_0 \tau\,, \tag{5.96b}$$

and the "time-time" linearized Einstein equation gives

$$\delta r_{\rm m}(\tau) = -\frac{2\alpha^2 C_1}{3\tau^3}\,. \tag{5.97}$$

This solution is complete, since the number of integration constants is four. The fourth integration constant C_4 implies $\delta v(\tau)/v(\tau)$ and $\delta\chi(\tau)/\chi(\tau)$ approach to nonzero constants in the limit of large time. However, $\delta h(\tau)/h(\tau)$ and $\delta r_{\rm m}(\tau)/r_{\rm m}(\tau)$ tend to zero as $1/\tau$, so that, in particular, the final value of $\tau^2 r_{\rm m}(\tau)$ is determined by the model parameters, rather than initial conditions in contrast to the case when $\tau\chi(\tau) \to 0$ for $\tau \to \infty$. This all means that values of v_0 and χ_0 are dependent of initial conditions, while

$$\lambda = -r_v(\tau) - r_\chi(\tau) = \lambda(v_0) + \lambda(\chi_0)\,, \tag{5.98}$$

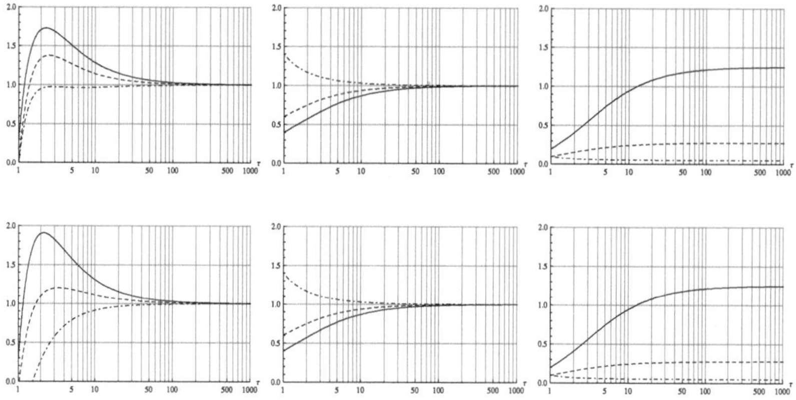

Figure 5.2: Numerical solutions of vector and Einstein field equations. The first column shows $-\lambda^{-1}(r_v(\tau) + r_\chi(\tau))$ as a function of rescaled time τ for three different sets of initial conditions. The second and third columns show $2\tau h(\tau)$ and $\tau^2 r_\mathrm{m}(\tau)$, respectively. The top and bottom lines differ from each other only by the values of the model parameters, namely the top line corresponds to $(\lambda,\, \zeta,\, \bar{\zeta},\, \zeta_S,\, \zeta_R) = (-1,\, -1,\, 0,\, 0,\, +1)$, while the bottom line – $(\lambda,\, \zeta,\, \bar{\zeta},\, \zeta_S,\, \zeta_R) = (+1,\, -1,\, 0,\, 0,\, -1)$. For both lines w_m equals $1/3$. Our numerical calculations are in complete agreement with (5.98) and (5.101), i.e. the value of $\tau^2 r_\mathrm{m}(\tau)$ at $\tau \to \infty$ depends on the initial conditions.

where $\lambda(v_0)$ and $\lambda(\chi_0)$ are given by (5.47) and (5.79), respectively, remains unchanged if (5.68) is valid (see Figure 5.1).

If $\bar{\zeta} = 0$, then $\zeta_S = 0$ and $\zeta_F = 0$ as these follow from (5.77), (5.72) and (5.69). These values of the coefficients correspond to $w_\mathrm{m} = 1/3$.(8)

In this case (5.71) and (5.75) become

$$(\dot{h} + 2h^2)\chi \;=\; 0\,. \tag{5.99}$$

In order to specify a solution, we must impose 5 initial conditions. This occurs, because (5.75) follows from (5.71), so that the constraint that the non-diagonal elements of the EMT of the vector field vanish is automatically valid. The equation (5.96b) as well as (5.96a) are still correct, where we should put $\alpha = -3/2$ and $\beta = 2\zeta_R/\zeta$, but (5.96b) and (5.97) become

$$\delta\chi(\tau) \;=\; \chi_0 C_1 + \frac{C_3}{\tau^{3/2}} - \frac{C_5}{3\zeta_R \chi_0 \tau} - \frac{9C_4}{4\zeta}(5\zeta + 6\zeta_R)v_0\tau \tag{5.100}$$

and

$$\delta r_\mathrm{m}(\tau) \;=\; \frac{C_5}{\tau^2} - \frac{3C_1}{2\tau^3}\,. \tag{5.101}$$

(8)This vector model was considered in [98]. There was made Ansatz $A_i = 0$. However, it seems artificial, so that, in what follows, we allow the differential equations to determine an evolution of $\chi(\tau)$ themselves.

Thus, in this case $h(\tau) \to 1/2\tau$ when τ approaches infinity, the final value of $\tau^2 r_{\rm m}(\tau)$ depends on the initial conditions, and (5.98) holds (see Figure 5.2).

5.3 General linear perturbations and Newton's law of gravity

As it was pointed out by Rubakov and Tinyakov in [91], the Dolgov model violates, in particular, the Newton gravitational law. The model under consideration, being a generalization of Dologv's one, suffers from the same malady.

To show this, let us consider general perturbations of both the vector and the metric fields:

$$\begin{aligned} A_\mu(t) &\to A_\mu(x) &=& A_\mu(t) + \delta A_\mu(x)\,, \\ g_{\mu\nu}(t) &\to g_{\mu\nu}(x) &=& g_{\mu\nu}(t) + \delta g_{\mu\nu}(x)\,, \end{aligned} \tag{5.102}$$

where δA_μ and δg_μ are small inhomogeneous perturbations under the background solution we have found above.

Our goal is to derive Newton's law of gravity in the hypothetical universe governed by (5.1). Therefore, we assume that timescales and wavelengths of the perturbations are extremely small in comparison with the universe age $H_0^{-1} \sim 10^{10}$ yr and the universe size $c/H_0 \sim 10^{26}$ m, respectively. In this case, as mentioned in Section 2.5, one can always consider a background that is roughly Minkowski, so that we take $\eta_{\mu\nu}$ instead of $g_{\mu\nu}(t)$ in (5.102).

Perturbation of vector field equation Linearizing the vector field equation (5.8) with respect to both δA_μ and $\delta g_{\mu\nu}$, one has

$$\bar\zeta \partial^2 \delta A_\mu + (2\zeta - \bar\zeta)\partial_\mu \partial^\lambda \delta A_\lambda = 2\zeta_R A_\mu \delta R - 2\zeta_S A^\lambda \delta R_{\mu\lambda} + 2\zeta A_\sigma \eta^{\lambda\rho} \delta\Gamma^\sigma_{\lambda\rho,\mu}\,, \tag{5.103}$$

where we have neglected terms $S\partial\delta g$, $\delta g \partial S$ and $F\partial\delta g$, $\delta g\partial F$, since they are much smaller at late time than $A\partial^2\delta g$ on the right-hand side (RHS) of (5.103).

Regarding the RHS of (5.103) as $\delta J_\mu(x)$, a solution of this equation can be represented as

$$\delta A_\mu(x) = \int d^4x' G_\mu{}^\nu(x,x')\delta J_\nu(x')\,, \tag{5.104}$$

where $G^\nu_\mu(x,x')$ is the Green function,

$$\left(\bar\zeta \eta_{\mu\lambda}\partial^2 + (2\zeta - \bar\zeta)\partial_\mu\partial_\lambda\right) G^{\lambda\nu}(x,x') = \delta^\nu_\mu\, \delta(x-x')\,. \tag{5.105}$$

A method of solving this equation is standard. First, it is clear from (5.105) that $G^{\mu\nu}(x,x') = G^{\mu\nu}(x-x')$. Second, one has to make the Fourier transformation of the Green function

$$G^{\mu\nu}(x-x') = \int \frac{d^4k}{(2\pi)^4}\, e^{ik(x-x')}\, \widetilde{G}^{\mu\nu}(k)\,. \tag{5.106}$$

Substituting this in (5.105), one obtains

$$-\left(\bar{\zeta}k^2\eta_{\mu\lambda}+(2\zeta-\bar{\zeta})k_\mu k_\lambda\right)\widetilde{G}^{\lambda\nu}(k) \;=\; \delta_\mu^\nu\,, \tag{5.107}$$

from which immediately follows

$$\widetilde{G}^{\mu\nu}(k) \;=\; -\frac{1}{\bar{\zeta}k^2}\left(\eta^{\mu\nu}+\frac{\bar{\zeta}-2\zeta}{2\zeta}\frac{k^\mu k^\nu}{k^2}\right), \tag{5.108}$$

assuming $\bar{\zeta}\neq 0$.(9) We have omitted a general solution of the homogeneous equation, i.e. (5.103) with the RHS set to zero, since we are mainly interested in how the RHS influences the evolution of δA_μ. We finally have

$$\delta A_\mu(x) \;=\; -\frac{1}{\bar{\zeta}}\int\frac{d^4k}{(2\pi)^4}\,e^{ikx}\,\frac{1}{k^2}\left(\delta_\mu^\nu+\frac{\bar{\zeta}-2\zeta}{2\zeta}\frac{k_\mu k^\nu}{k^2}\right)\delta\widetilde{J}_\nu(k)\,. \tag{5.109}$$

where $\delta\widetilde{J}_\nu(k)$ is the Fourier transform of

$$\delta J_\mu(x) \;\equiv\; 2\zeta_R A_\mu\delta R-2\zeta_S A^\lambda\delta R_{\mu\lambda}+2\zeta A_\sigma\eta^{\lambda\rho}\delta\Gamma^\sigma_{\lambda\rho,\mu}\,. \tag{5.110}$$

Perturbation of vector energy-momentum tensor The linear perturbation of the vector energy-momentum tensor is

$$\begin{aligned}\delta T^{\mathrm{A}}_{\mu\nu}(x) \;=\;& \zeta_S\Big(A_\mu\Big[\partial^2\delta A_\nu+\partial_\nu\partial^\lambda\delta A_\lambda-2A_\sigma\partial^\lambda\delta\Gamma^\sigma_{\nu\lambda}\Big]+A_\nu\Big[\partial^2\delta A_\mu+\partial_\mu\partial^\lambda\delta A_\lambda\\&-2A_\sigma\partial^\lambda\delta\Gamma^\sigma_{\mu\lambda}\Big]-A_\lambda\Big[\partial^\lambda\partial_\mu\delta A_\nu+\partial^\lambda\partial_\nu\delta A_\mu-2A_\sigma\partial^\lambda\delta\Gamma^\sigma_{\mu\nu}\Big]\Big)\\&+2\zeta_Q\left(A_\mu\partial_\nu+A_\nu\partial_\mu-\eta_{\mu\nu}A^\kappa\partial_\kappa\right)\left(\partial^\lambda\delta A_\lambda-A_\sigma\eta^{\lambda\rho}\delta\Gamma^\sigma_{\lambda\rho}\right)\\&+\zeta_R\Big(\Big[A^2\eta_{\mu\nu}-2A_\mu A_\nu\Big]\delta R-2A^2\delta R_{\mu\nu}+2L_{\mu\nu}\Big[2A^\lambda\delta A_\lambda+A_\lambda A_\rho\delta g^{\lambda\rho}\Big]\Big)\,,\end{aligned} \tag{5.111}$$

where $L_{\mu\nu}=\partial_\mu\partial_\nu-\eta_{\mu\nu}\partial^2$ (see Appendix D for details). We have taken into account only terms which depend on the highest (second) derivative of the perturbations as the most relevant for Newton's gravity.

Making the Fourier transformation of $\delta T_{\mu\nu}(x)$, we obtain

$$\begin{aligned}\delta\widetilde{T}^{\mathrm{A}}_{\mu\nu}(k) \;=\;& \Big(k^\sigma\Big[\zeta_S M^{\lambda\rho}_{\mu\nu\sigma}+\zeta_Q N^{\lambda\rho}_{\mu\nu\sigma}+\zeta_R K^{\lambda\rho}_{\mu\nu\sigma}\Big]+2\zeta_R\Big[k^2\eta_{\mu\nu}-k_\mu k_\nu\Big]A^\lambda A^\rho\Big)\delta\widetilde{g}_{\lambda\rho}\\&-2k_\rho\eta^{\sigma\kappa}\Big(\zeta_S M^{\lambda\rho}_{\mu\nu\sigma}+\zeta_Q N^{\lambda\rho}_{\mu\nu\sigma}+\zeta_R K^{\lambda\rho}_{\mu\nu\sigma}\Big)\delta\widetilde{g}_{\lambda\kappa}\,,\end{aligned} \tag{5.112}$$

where $M^{\lambda\rho}_{\mu\nu\sigma}$, $N^{\lambda\rho}_{\mu\nu\sigma}$ and $K^{\lambda\rho}_{\mu\nu\sigma}$ can be found in Appendix D.

(9) If $\bar{\zeta}=0$, then ζ_F and ζ_S must be zero as well. In this case, $\mathcal{L}_{\mathrm{A}}=\zeta_Q(\nabla^\mu A_\mu)^2+\zeta_R RA^2$ and looks quite similar to the scalar model considered in Section 4.2. Therefore, the effective gravitational constant G_{eff} in this vector model has an analogous form to (4.13), where we must replace ϕ^2 by A^2. Consequently, we can rule out this vector-tensor theory (5.2) with $\zeta_F=\zeta_S=0$ by applying the same argument as in Section 4.2.

We see that there appear terms in the perturbed Einstein equations which depend on the second derivative of the metric perturbation multiplied by t^2. For example, in a symbolic notation $M \sim A^2k^2$ and the same for N and K, what results in $\delta T \sim A^2\partial^2\delta g \sim t^2\partial^2\delta g$. Clearly, these terms modify the Newton gravity law and the behavior of gravitational waves in an unacceptable way.

Let us suppose one can choose the Lagrangian parameters ζ_f, $f \in \{S, F, Q, R\}$ in such a way, that these bad terms are canceled out. Excluding the trivial choice $\zeta_f = 0$, one can easily see that if we put $\zeta_S M^{\lambda\rho}_{\mu\nu\sigma} + \zeta_Q N^{\lambda\rho}_{\mu\nu\sigma} + \zeta_R K^{\lambda\rho}_{\mu\nu\sigma} = 0$, then there still remains a bad term in $\delta\widetilde{T}_{\mu\nu}$, namely $2\zeta_R(k^2\eta_{\mu\nu} - k_\mu k_\nu)A^\lambda A^\rho \delta\widetilde{g}_{\lambda\rho}$. Therefore, one has to set $\zeta_R = 0$. Hence, one needs the following equation to be satisfied

$$\zeta_S M^{\lambda\rho}_{\mu\nu\sigma} + \zeta_Q N^{\lambda\rho}_{\mu\nu\sigma} = 0\,. \tag{5.113}$$

Comparing terms in $M^{\lambda\rho}_{\mu\nu\sigma}$ and $N^{\lambda\rho}_{\mu\nu\sigma}$ with each other, one arrives at a conclusion that this is possible only if both ζ_S and ζ_Q are zero. However, we have excluded this case. Consequently, this vector model is in a contradiction with Newton's gravity.

Chapter 6

Vector-tensor model II

Let us consider a vector-tensor model with two vector fields, A_μ and B_μ, the dynamics of which are determined by the following action functional [101, 102, 103]

$$S[g, A, B] \;=\; -\int d^4x\sqrt{-g}\,\epsilon(\mathcal{L}_{\rm A}, \mathcal{L}_{\rm B})\,, \tag{6.1}$$

where $\epsilon(\mathcal{L}_{\rm A}, \mathcal{L}_{\rm B})$ is given by

$$\epsilon(\mathcal{L}_{\rm A}, \mathcal{L}_{\rm B}) \;=\; a\,\frac{\mathcal{L}_{\rm A}}{\mathcal{L}_{\rm B}} + b\,\frac{\mathcal{L}_{\rm B}}{\mathcal{L}_{\rm A}}\,. \tag{6.2}$$

Here a and b are real numbers ($|a|+|b| \neq 0$), which values we do not specify in order to obtain a general result, $\mathcal{L}_{\rm A}$ is completely identical to (5.4) and $\mathcal{L}_{\rm B}$ is $\mathcal{L}_{\rm A}$ with A_μ replaced by B_μ. Thus the coefficients (5.5) are assumed to be the same for both vector fields.

The ϵ-function is taken to be (6.2), because then it possesses the following properties

$$\mathcal{L}_{\rm A}\frac{\partial\epsilon}{\partial\mathcal{L}_{\rm A}} + \mathcal{L}_{\rm B}\frac{\partial\epsilon}{\partial\mathcal{L}_{\rm B}} \;=\; 0\,, \tag{6.3a}$$

$$\mathcal{L}_{\rm A}^2\frac{\partial^2\epsilon}{\partial\mathcal{L}_{\rm A}^2} + 2\mathcal{L}_{\rm A}\mathcal{L}_{\rm B}\frac{\partial^2\epsilon}{\partial\mathcal{L}_{\rm A}\partial\mathcal{L}_{\rm B}} + \mathcal{L}_{\rm B}^2\frac{\partial^2\epsilon}{\partial\mathcal{L}_{\rm B}^2} \;=\; 0 \tag{6.3b}$$

and so on, which turn out to be crucial for the preservation of the Newton gravity law as we will see shortly. In principle, there are infinitely many such functions.[(1)] However, it seems (6.2) is the most simple one and this motivates our choice.

[(1)]In general, any function $f(x, y) = f(x/y)$ satisfies (6.3). Indeed, first, if $f(x, y)$ satisfies (6.3a), then (6.3b) is also satisfied. Therefore, it is sufficient to find all functions which are solutions of the equation: $x\partial_x f + y\partial_y f = 0$, where $\partial_x \equiv \partial/\partial x$ and $\partial_y \equiv \partial/\partial y$. Second, this equation can be rewritten as follows: $(\partial_q - \partial_p)f(q, p) = 0$, where q and p have been defined as $q = \ln(x)$ and $p = -\ln(y)$, respectively. A solution of this equation can be straightforwardly found and is $f(q, p) = f(q + p)$, from which immediately follows $f(x, y) = f(x/y)$ which was to be proved.

6.1 Vector and Einstein field equations

The vector field equations for both A_μ and B_μ can be obtained from the model with one vector field, which we have considered in the previous chapter, by simply making a change

$$\zeta_f \;\to\; \frac{\partial \epsilon}{\partial \mathcal{L}_{\rm V}}\zeta_f\,, \quad \text{for} \quad f \;\in\; \{S, F, Q, R\}\,, \tag{6.4}$$

where $V_\mu \in \{A_\mu, B_\mu\}$. This procedure gives

$$\begin{aligned} \bar\zeta_{\,\rm A} F^{\lambda}{}_{\mu;\lambda} + \zeta_{\,\rm A} S^{\lambda}_{\lambda;\mu} &= 2\zeta_R R A_\mu - 2\zeta_S R_{\mu\lambda} A^\lambda \\ &\quad -\big(\zeta_{S\,\rm A} S^{\lambda}_{\mu} + \zeta_{F\,\rm A} F^{\lambda}{}_{\mu} + \zeta_{Q\,\rm A} S^{\rho}_{\rho}\delta^{\lambda}_{\mu}\big)\partial_\lambda \ln \big|\,\epsilon'_{\mathcal{L}_{\rm A}}\big|\,, \end{aligned} \tag{6.5a}$$

$$\begin{aligned} \bar\zeta_{\,\rm B} F^{\lambda}{}_{\mu;\lambda} + \zeta_{\,\rm B} S^{\lambda}_{\lambda;\mu} &= 2\zeta_R R B_\mu - 2\zeta_S R_{\mu\lambda} B^\lambda \\ &\quad -\big(\zeta_{S\,\rm B} S^{\lambda}_{\mu} + \zeta_{F\,\rm B} F^{\lambda}{}_{\mu} + \zeta_{Q\,\rm B} S^{\rho}_{\rho}\delta^{\lambda}_{\mu}\big)\partial_\lambda \ln \big|\,\epsilon'_{\mathcal{L}_{\rm B}}\big|\,, \end{aligned} \tag{6.5b}$$

where by definition $\epsilon'_{\mathcal{L}_{\rm A}}$ and $\epsilon'_{\mathcal{L}_{\rm B}}$ are partial derivatives of $\epsilon(\mathcal{L}_{\rm A}, \mathcal{L}_{\rm B})$ with respect to $\mathcal{L}_{\rm A}$ and $\mathcal{L}_{\rm B}$, respectively. As one can see there appear extra terms in the vector field equations (6.5) in comparison with (5.8). It occurs due to a non-quadratic dependence of (6.2) on its variables.

After variation of the action of the whole system with respect to the metric field $g_{\mu\nu}$, we obtain the Einstein equations which read

$$R_{\mu\nu} - \frac{1}{2} R\, g_{\mu\nu} \;=\; 8\pi G\Big(\rho_\Lambda g_{\mu\nu} + T^{\rm vec}_{\mu\nu} + T^{\rm m}_{\mu\nu}\Big)\,, \tag{6.6}$$

where $T^{\rm m}_{\mu\nu}$ is the energy-momentum tensor of the matter field ψ regarded as a perfect fluid and, consequently, $T^{\rm m}_{\mu\nu}$ takes the form (2.10), $T^{\rm vec}_{\mu\nu}$ is the energy-momentum tensor of the vector fields coming from the variation of (6.1) over the metric

$$\begin{aligned} T^{\rm vec}_{\mu\nu} &= \left(\epsilon - \mathcal{L}_{\rm A}\frac{\partial \epsilon}{\partial \mathcal{L}_{\rm A}} - \mathcal{L}_{\rm B}\frac{\partial \epsilon}{\partial \mathcal{L}_{\rm B}}\right) g_{\mu\nu} \\ &\quad + \sum_{\rm V} \epsilon'_{\mathcal{L}_{\rm V}}\Big(T^{S}_{\mu\nu}(V) + T^{F}_{\mu\nu}(V) + T^{Q}_{\mu\nu}(V) + T^{R}_{\mu\nu}(V) + T^{\Delta}_{\mu\nu}(V)\Big)\,, \end{aligned} \tag{6.7}$$

where $T^{f}_{\mu\nu}$, $f \in \{S, F, Q, R\}$ as in (5.10), but the last term $T^{\Delta}_{\mu\nu}$ comes from the non-quadratic dependence of the ϵ-function on its variables

$$\begin{aligned} T^{\Delta}_{\mu\nu}(V) &= \Big(\zeta_S\big[2\,{}_{\rm V}S^{\lambda}_{(\mu}V_{\nu)} - {}_{\rm V}S_{\mu\nu}V^\lambda\big] + \zeta_{Q}\,{}_{\rm V}S^{\rho}_{\rho}\big[2V_{(\mu}\delta^{\lambda}_{\nu)} - g_{\mu\nu}V^\lambda\big]\Big)\partial_\lambda \ln \big|\,\epsilon'_{\mathcal{L}_{\rm V}}\big| \\ &\quad + 2\zeta_R\left(\epsilon'_{\mathcal{L}_{\rm V}}\right)^{-1}\Big(L_{\mu\nu}\left(\epsilon'_{\mathcal{L}_{\rm V}}V^2\right) - \epsilon'_{\mathcal{L}_{\rm V}}L_{\mu\nu}(V^2)\Big) \end{aligned} \tag{6.8}$$

(see Appendix E). Note, the EMT of the vector fields (6.7) is general, i.e. it has this form for any ϵ-function. For example, if we take $\epsilon(\mathcal{L}_{\rm A}, \mathcal{L}_{\rm B}) = \mathcal{L}_{\rm A}$, we obtain (5.10). In our special case of the ϵ-function, in particular, the second and third terms in the first round brackets on the right-hand side of (6.7) cancel each other.

6.2 Flat, homogeneous and isotropic universe

Vector and Einstein field equations

For the flat Friedmann-Robertson-Walker metric tensor (5.11) and homogeneous vector fields, $A_\mu(x) = A_\mu(t)$ and $B_\mu(x) = B_\mu(t)$, the vector field equations (6.5a) and (6.5b) become in components

$$\ddot{v}_{\rm A} + 3h\dot{v}_{\rm A} + 3\left(\dot{h} + \zeta^{-1}\Big[(2\zeta_R - \zeta_S)\dot{h} + (4\zeta_R - \zeta_S)h^2\Big]\right) v_{\rm A} + \left(\dot{v}_{\rm A} + 3(\zeta_Q/\zeta)hv_{\rm A}\right)\partial_0 \ln|\epsilon'_{\mathcal{L}_{\rm A}}| = 0\,, \tag{6.9a}$$

$$\ddot{v}_{\rm B} + 3h\dot{v}_{\rm B} + 3\left(\dot{h} + \zeta^{-1}\Big[(2\zeta_R - \zeta_S)\dot{h} + (4\zeta_R - \zeta_S)h^2\Big]\right) v_{\rm B} + \left(\dot{v}_{\rm B} + 3(\zeta_Q/\zeta)hv_{\rm B}\right)\partial_0 \ln|\epsilon'_{\mathcal{L}_{\rm B}}| = 0\,, \tag{6.9b}$$

where we have introduced the dimensional variables as in (5.15), and

$$\ddot{\chi}_{\rm A} + 3h\dot{\chi}_{\rm A} + \left(\dot{h} + 2h^2 + 2\bar{\zeta}^{-1}\Big[(6\zeta_R - \zeta_S)\dot{h} + 3(4\zeta_R - \zeta_S)h^2\Big]\right)\chi_{\rm A} + \left(\dot{\chi}_{\rm A} + (1 - 2\zeta_S/\bar{\zeta})\,h\chi_{\rm A}\right)\partial_0 \ln|\epsilon'_{\mathcal{L}_{\rm A}}| = 0\,, \tag{6.10a}$$

$$\ddot{\chi}_{\rm B} + 3h\dot{\chi}_{\rm B} + \left(\dot{h} + 2h^2 + 2\bar{\zeta}^{-1}\Big[(6\zeta_R - \zeta_S)\dot{h} + 3(4\zeta_R - \zeta_S)h^2\Big]\right)\chi_{\rm B} + \left(\dot{\chi}_{\rm B} + (1 - 2\zeta_S/\bar{\zeta})\,h\chi_{\rm B}\right)\partial_0 \ln|\epsilon'_{\mathcal{L}_{\rm B}}| = 0\,. \tag{6.10b}$$

Here we have assumed that $A_i(t)$ and $B_i(t)$ have the following structure

$$A_i(\tau) \equiv M_{\rm Planck}\, a(\tau)\,\chi_{\rm A}(\tau)\,\xi_i^{\rm A}\,, \quad B_i(\tau) \equiv M_{\rm Planck}\, a(\tau)\,\chi_{\rm B}(\tau)\,\xi_i^{\rm B}\,, \tag{6.11}$$

where $a(\tau)$ is the scale factor expressed via rescaled cosmic time τ, $\xi_i^{\rm A}$ and $\xi_i^{\rm B}$ are unit constant three-dimensional vectors.

Equations (6.9) and (6.10) are supplemented by the Einstein equations

$$3h^2 = \lambda + r_{\rm vec} + r_{\rm m}\,, \tag{6.12a}$$

$$2\dot{h} + 3h^2 = \lambda - p_{\rm vec} - w_{\rm m} r_{\rm m} \tag{6.12b}$$

as well as the equation describing the evolution of the perfect fluid (5.13d), namely

$$\dot{r}_{\rm m} + 3h(1 + w_{\rm m})r_{\rm m} = 0\,, \tag{6.13}$$

and the condition that the non-diagonal elements of $T^{\rm vec}_{\mu\nu}$ vanish. The explicit form for the rescaled energy density $r_{\rm vec}(\tau)$ and pressure $p_{\rm vec}(\tau)$ and the non-diagonal elements of $T^{\rm vec}_{\mu\nu}$ can be found with the aid of Appendix C and Appendix E.

A particular class of exact solutions

The vector and Einstein field equations are nonlinear, therefore it is hardly possible to find their full exact solution. However, there exist the following particular exact solution:

$$v_{\rm A}(\tau) = C_1(\tau-\tau_0)^x, \quad \chi_{\rm A}(\tau) = 0, \tag{6.14a}$$

$$v_{\rm B}(\tau) = C_2(\tau-\tau_0)^x, \quad \chi_{\rm B}(\tau) = 0, \tag{6.14b}$$

$$h(\tau) = \frac{y}{\tau-\tau_0}, \tag{6.14c}$$

$$r_{\rm m}(\tau) = \frac{3y^2}{(\tau-\tau_0)^2}, \tag{6.14d}$$

where C_1, C_2 and τ_0 are some constants, x and y satisfy an equation

$$\alpha(x-1)(x+3y)+3\beta\Big(2(2\alpha+3)x+3(y-\alpha-2)\Big)y = 0 \tag{6.15}$$

depending on the parameters of the Lagrangian via α and β defined in (5.50). Moreover, we must have

$$w_{\rm m}+1-\frac{2}{3y} = 0, \tag{6.16a}$$

$$\lambda+a\frac{C_1^2}{C_2^2}+b\frac{C_2^2}{C_1^2} = 0. \tag{6.16b}$$

We see that a ratio C_1/C_2 is fixed by the value of the rescaled cosmological constant λ. In other words, (6.14) depends on two arbitrary constants: τ_0 and C_1C_2, while, generally speaking, there must be seven integration constants.[(2)] Actually, we have to impose $7+2+2$ initial conditions if we take into account $\chi_{\rm A}(\tau)$ and $\chi_{\rm B}(\tau)$. For the moment, we do not consider the spatial components of the vector fields as being dynamical, so that we talk only about seven initial conditions.

Note that y is fixed by the matter equation of state parameter $w_{\rm m}$. This situation is contrary to that we had in Chapter 5. Here the evolution of the universe is determined by the matter field, rather than by the ζ-coefficients. However, x still depends on them and on $w_{\rm m}$ as this follows from (6.15).

Stability analysis In order to analyze whether the found class of the exact solutions is asymptotically stable or not, we consider homogeneous perturbations of the variables around (6.14). In what follows, we omit τ_0 in (6.14) for the sake of simplicity. It does not make further treatment less general, since τ_0 reveals itself in the perturbations as we will see this shortly.

Since $\mathcal{L}_{\rm A}$ depends on $\chi_{\rm A}(\tau)$ quadratically, i.e. it is a function of $\chi_{\rm A}^2$, $\chi_{\rm A}\dot\chi_{\rm A}$ and $\dot\chi_{\rm A}^2$, and we have taken $\chi_{\rm A}(\tau)=0$, there are no terms in the linearized (6.9) and (6.12) depending on $\delta\chi_{\rm A}(\tau)$. Similarly,

[(2)]For two specific choices of the ζ-coefficients, there are particular exact solutions with four arbitrary constants: 1) $\zeta_S = 0$: $v_{\rm A}(\tau) = C_1(\tau-\tau_0)+(C_3/C_2)(\tau-\tau_0)^{-3y}$, $v_{\rm B}(\tau) = C_2(\tau-\tau_0)+(C_4/C_1)(\tau-\tau_0)^{-3y}$ and $\chi_{\rm A}(\tau)=\chi_{\rm B}(\tau)=0$ with $C_4=\pm C_3$ as well as (6.14c), (6.14d), (6.16), where $y=1/2$; 2) $\zeta_S=\zeta_R=0$: $v_{\rm A}(\tau)=C_1(\tau-\tau_0)+C_3(\tau-\tau_0)^{-3y}$, $v_{\rm B}(\tau)=C_2(\tau-\tau_0)+C_4(\tau-\tau_0)^{-3y}$, $\chi_{\rm A}(\tau)=\chi_{\rm B}(\tau)=0$ with (6.14c), (6.14d) and (6.16).

these equations do not have terms with $\delta\chi_{\rm B}(\tau)$. Therefore, we are able to consider $\delta\chi_{\rm A}(\tau)$ and $\delta\chi_{\rm B}(\tau)$ separately.

Linearized (6.9)**,** (6.12a) **and** (6.13) **and their solution** Keeping only terms linear with respect to the homogeneous perturbations $\delta v_{\rm A}(\tau)$, $\delta v_{\rm B}(\tau)$ and $\delta h(\tau)$, we find

$$\frac{1}{C_1}\left(\pi_1^a\,\delta\ddot{v}_{\rm A}+\frac{\pi_2^a}{\tau}\,\delta\dot{v}_{\rm A}+\frac{\pi_3^a}{\tau^2}\,\delta v_{\rm A}\right)+\frac{1}{C_2}\left(\pi_1^c\,\delta\ddot{v}_{\rm B}+\frac{\pi_2^c}{\tau}\,\delta\dot{v}_{\rm B}+\frac{\pi_3^c}{\tau^2}\,\delta v_{\rm B}\right)$$
$$+\,\tau^{x+1}\left(\pi_1^h\,\delta\ddot{h}+\frac{\pi_2^h}{\tau}\,\delta\dot{h}+\frac{\pi_3^h}{\tau^2}\,\delta h\right)\;=\;0\,, \tag{6.17a}$$

$$\frac{1}{C_2}\left(\pi_1^b\,\delta\ddot{v}_{\rm B}+\frac{\pi_2^b}{\tau}\,\delta\dot{v}_{\rm B}+\frac{\pi_3^b}{\tau^2}\,\delta v_{\rm B}\right)+\frac{1}{C_1}\left(\pi_1^c\,\delta\ddot{v}_{\rm A}+\frac{\pi_2^c}{\tau}\,\delta\dot{v}_{\rm A}+\frac{\pi_3^c}{\tau^2}\,\delta v_{\rm A}\right)$$
$$-\,\tau^{x+1}\left(\pi_1^h\,\delta\ddot{h}+\frac{\pi_2^h}{\tau}\,\delta\dot{h}+\frac{\pi_3^h}{\tau^2}\,\delta h\right)\;=\;0\,. \tag{6.17b}$$

The coefficients π_i^a, π_i^b, π_i^c and π_i^h ($i=1,2,3$) depend only on the constant parameters of the Lagrangian density, namely ζ_f, $f\in\{S,F,Q,R\}$ as well as $w_{\rm m}$ and can be found in Appendix F.

The "time-time" Einstein equation (6.12a) and the equation describing the evolution of the perfect fluid (6.13) linearized with respect to $\delta v_{\rm A}(\tau)$, $\delta v_{\rm B}(\tau)$, $\delta h(\tau)$ and $\delta r_{\rm m}(\tau)$ are

$$\frac{6y}{\tau}\,\delta h-\delta r_{\rm vec}-\delta r_{\rm m}\;=\;0\,, \tag{6.18a}$$

$$\delta\dot{r}_{\rm m}+\frac{2}{\tau}\,\delta r_{\rm m}+\frac{6y}{\tau^2}\,\delta h\;=\;0\,, \tag{6.18b}$$

where the linearized energy density of the vector fields is given by

$$\delta r_{\rm vec}=\frac{\tau^{2-x}}{C_1}\left(\pi_1^r\delta\ddot{v}_{\rm A}+\frac{\pi_2^r}{\tau}\delta\dot{v}_{\rm A}+\frac{\pi_3^r}{\tau^2}\delta v_{\rm A}\right)-\frac{\tau^{2-x}}{C_2}\left(\pi_1^r\delta\ddot{v}_{\rm B}+\frac{\pi_2^r}{\tau}\delta\dot{v}_{\rm B}+\frac{\pi_3^r}{\tau^2}\delta v_{\rm B}\right) \tag{6.19}$$

with coefficients π_i^r ($i=1,2,3$) depending only on ζ_f, $f\in\{S,F,Q,R\}$ and $w_{\rm m}$. They can be found in Appendix F as well.

It is straightforward to show that $\delta r_{\rm vec}(\tau)$ can be expressed only in terms of the following combination of $\delta v_{\rm A}(\tau)$ and $\delta v_{\rm B}(\tau)$:

$$\delta v_{\rm C}(\tau)\;\equiv\;\frac{1}{C_1}\,\delta v_{\rm A}(\tau)-\frac{1}{C_2}\,\delta v_{\rm B}(\tau)\,. \tag{6.20}$$

Moreover, it turns out one can obtain a differential equation only on $\delta v_{\rm C}(\tau)$ from (6.17a) and (6.17b).(3) This is achieved by subtracting (6.17b) from (6.17a) and taking into account that equalities $\pi_i^a+\pi_i^b+2\pi_i^c=0$ for $i=1,2,3$ hold. So we have

$$\left(\pi_1^a+\pi_1^c\right)\delta\ddot{v}_{\rm C}+\frac{\pi_2^a+\pi_2^c}{\tau}\,\delta\dot{v}_{\rm C}+\frac{\pi_3^a+\pi_3^c}{\tau^2}\,\delta v_{\rm C}\;=\;0\,. \tag{6.21}$$

(3)Shortly we will see that it occurs due to the special properties of the ϵ-function (6.3) as well as absence of the time derivatives of δh in (6.19).

Assuming $\pi_1^a + \pi_1^c \neq 0$ (otherwise (6.21) becomes a trivial identity: $0 = 0$), it is easy to show that

$$\delta v_{\rm C}(\tau) \;=\; \widetilde{C}_1 \tau^x + \widetilde{C}_2 \tau^{x-3y-1} \tag{6.22}$$

is an exact full solution of (6.21). $\widetilde{C}_{1,2}$ are integration constants. Now substituting $\delta v_{\rm C}$ in $\delta r_{\rm vec}(\tau)$, we obtain

$$\delta r_{\rm vec}(\tau) \;=\; 2\widetilde{C}_1 \left(a\frac{C_1^2}{C_2^2} - b\frac{C_2^2}{C_1^2} \right), \tag{6.23}$$

i.e. $\delta r_{\rm vec}(\tau)$ is a constant specified by imposing initial conditions. Consequently, in order to satisfy the linearized "time-time" Einstein equation (6.18a), $h(\tau)\delta h(\tau)$ and $\delta r_{\rm m}(\tau)$ must be constant as well. In other words, $\delta h(\tau) \sim \tau \delta r_{\rm vec}(\tau)$ and $\delta r_{\rm m}(\tau) \sim \delta r_{\rm vec}(\tau)$, i.e.

$$\delta h(\tau)/h(\tau) \;\sim\; \widetilde{C}_1 \tau^2, \quad \delta r_{\rm m}(\tau)/r_{\rm m}(\tau) \;\sim\; \widetilde{C}_1 \tau^2 \tag{6.24}$$

for large time. This situation is quite similar to that we encountered in Subsection 5.2.1. It means the background solution (6.14) is unstable, unless one imposes initial conditions in a suitable manner, namely such that $\widetilde{C}_1 = 0$. In other words, one has to remove a solution which makes instability. If so, then we are left with the second independent solution of $\delta v_{\rm C}(\tau)$ that has no contribution to the homogeneous perturbation of the energy density $\delta r_{\rm vec}(\tau)$. Solving (6.18a) and (6.18b), we obtain

$$\delta h(\tau) \;=\; -\frac{y\widetilde{C}_3}{x\tau^2}, \quad \delta r_{\rm m}(\tau) \;=\; -\frac{6y^2\widetilde{C}_3}{x\tau^3}. \tag{6.25}$$

Adding (6.17b) to (6.17a) and using explicit expressions of $\delta v_{\rm C}(\tau)$ and $\delta h(\tau)$, one finds

$$(\pi_1^a - \pi_1^b)\,\delta\ddot{v}_{\rm D} + \frac{\pi_2^a - \pi_2^b}{\tau}\,\delta\dot{v}_{\rm D} + \frac{\pi_3^a - \pi_3^b}{\tau^2}\,\delta v_{\rm D} \;=\; \frac{4y\widetilde{C}_3}{x\tau^{3-x}}\left(6\pi_1^h - 2\pi_2^h + \pi_3^h\right), \tag{6.26}$$

where by definition

$$\delta v_{\rm D}(\tau) \;\equiv\; \frac{1}{C_1}\,\delta v_{\rm A}(\tau) + \frac{1}{C_2}\,\delta v_{\rm B}(\tau). \tag{6.27}$$

The equation (6.26) has two independent solutions or, in other words, two arbitrary constants of integration. As mentioned above, τ_0 must reveal itself in the homogeneous perturbations. In order to see how this occurs, consider (6.14) at $\tau \gg \tau_0$. Then we are allowed to expand (6.14) in terms of τ_0/τ in this limit: $v_{\rm A}(\tau) = C_1(\tau - \tau_0)^x \approx C_1\tau^x - x\tau_0 C_1 \tau^{x-1}$ and the same for $v_{\rm B}(\tau)$ with C_2 instead of C_1. Besides, since $h(\tau) = y/(\tau - \tau_0) \approx y/\tau + y\tau_0/\tau^2$, we find $\tau_0 = -\widetilde{C}_3/x$. Hence, there must be a solution of (6.26) such as $\delta v_{\rm D}(\tau) = 2\widetilde{C}_3\tau^{x-1}$. Direct calculation shows that is indeed so. This is simply a particular solution of inhomogeneous differential equation (6.26). We are left to find two independent solutions of the homogeneous (6.26), i.e. (6.26) with omitted right-hand side. Having solved this, we obtain

$$C_1^{-1}\delta v_{\rm A}(\tau) \;=\; \widetilde{C}_3\tau^{x-1} + \widetilde{C}_4\tau^x + \left(\widetilde{C}_5 + \frac{1}{2}\widetilde{C}_2\right)\tau^{x-3y-1}, \tag{6.28a}$$

$$C_2^{-1}\delta v_{\rm B}(\tau) \;=\; \widetilde{C}_3\tau^{x-1} + \widetilde{C}_4\tau^x + \left(\widetilde{C}_5 - \frac{1}{2}\widetilde{C}_2\right)\tau^{x-3y-1}. \tag{6.28b}$$

The number of integration constants is 5, i.e. $\widetilde{C}_1 = 0$ and $\widetilde{C}_{2,\ldots,5}$, this coincides with the number of integration constants of the linearized vector and Einstein equations,[(4)] i.e. (6.25) and (6.28) give a full solution of (6.17a), (6.17b), (6.18a) and (6.18b).

If $x - 3y - 1$ is less than $x - 1$ or if y is positive, then (6.14) is asymptotically stable subject to the exclusion of the solution with nonzero $\widetilde{C}_1$ by imposing initial conditions in the proper way. The question is now how $\delta\chi_{\rm A}(\tau)$ and $\delta\chi_{\rm B}(\tau)$ evolve with time around $\chi_{\rm A}(\tau) = \chi_{\rm B}(\tau) = 0$.

Linearized (6.10a), (6.10b) **and their solution** Since both $\delta\chi_{\rm A}(\tau)$ and $\delta\chi_{\rm B}(\tau)$ satisfy the same differential equation, it is sufficient to consider one of them only, say, $\delta\chi_{\rm A}(\tau)$. Linearizing (6.10a) with respect to $\delta\chi_{\rm A}(\tau)$, one obtains

$$\delta\ddot{\chi}_{\rm A} + \frac{\pi_1^\chi}{\tau}\,\delta\dot{\chi}_{\rm A} + \frac{\pi_2^\chi}{\tau^2}\,\delta\chi_{\rm A} \;=\; 0\,, \tag{6.29}$$

where $\pi_{1,2}^\chi$ do not depend on rescaled cosmic time τ and can be found in Appendix F. Looking for a solution in a form $\delta\chi_{\rm A}(\tau) \sim \tau^z$, we find

$$z^\pm \;=\; \frac{1}{2}\Big(1 - \pi_1^\chi \pm \sqrt{\left(1-\pi_1^\chi\right)^2 - 4\pi_2^\chi}\Big)\,. \tag{6.30}$$

If both z^+ and z^- are negative, then $\delta\chi_{\rm A}(\tau)$ and $\delta\chi_{\rm B}(\tau)$ decrease sufficiently fast with growing time, i.e. they do not spoil the background solution.[(5)]

6.3 General linear perturbations and Newton's law of gravity

Let us consider general perturbations of the fields under consideration:

$$\begin{aligned}
A_\mu(t) &\;\to\; A_\mu(x) &=&\; A_\mu(t) \;+\; \delta A_\mu(x)\,,\\
B_\mu(t) &\;\to\; B_\mu(x) &=&\; B_\mu(t) \;+\; \delta B_\mu(x)\,,\\
g_{\mu\nu}(t) &\;\to\; g_{\mu\nu}(x) &=&\; g_{\mu\nu}(t) \;+\; \delta g_{\mu\nu}(x)\,,
\end{aligned} \tag{6.31}$$

where $A_\mu(t)$, $B_\mu(t)$ and $g_{\mu\nu}(t)$ on the right-hand side of (6.31) are the background solution found above. In what follows, we assume $|\delta A(x)| \ll |A(t)|$, $|\delta B(x)| \ll |B(t)|$ and $|\delta g(x)| \ll |g(t)|$, so that we will take into account in the vector and Einstein field equations only terms depending linearly on the perturbations.

---

(4) Indeed, if we set $\delta v_{\rm A}(\tau)$, $\delta\dot{v}_{\rm A}(\tau)$, $\delta v_{\rm B}(\tau)$ and $\delta\dot{v}_{\rm B}(\tau)$ at certain initial moment of time, say, $\tau_{\rm in}(\tau)$, then we know $\delta v_{\rm C}(\tau)$, $\delta\dot{v}_{\rm C}(\tau)$ and $\delta\ddot{v}_{\rm C}(\tau)$ at that time as it follows from (6.21), provided $C_{1,2}$ are known. Now if we also set $\delta r_{\rm m}(\tau)$ at $\tau_{\rm in}$, then we are able to find $\delta h(\tau)$, $\delta\dot{h}(\tau)$ and $\delta\ddot{h}(\tau)$ at this time by use of (6.18a) and (6.18b). So that we need only to fix 5 integration constants. Note, this occurs due to the fact that (6.3) and $v_{\rm A}/C_1 - v_{\rm B}/C_2 = 0$. If, in general, we could find another class of exact solutions, such that $v_{\rm A}/C_1 - v_{\rm B}/C_2 \neq 0$, then there would be terms in (6.19) with $\delta h(\tau)$ and its first and second derivatives. As a consequence, we would have to impose initial values of $\delta h(\tau)$ and $\delta\dot{h}(\tau)$ at $\tau_{\rm in}$ as well.

(5) See discussion in Subsection 5.2.3.

Perturbation of the vector field equations

Linearizing vector field equations (6.5a) and (6.5b) around the background solution, we obtain

$$\begin{aligned}
&\bar{\zeta}\delta\big({}_{\mathrm{A}}F^{\lambda}{}_{\mu;\lambda}\big) + \zeta\delta\big({}_{\mathrm{A}}S^{\lambda}_{\lambda;\mu}\big) + 2\zeta_S\big(R_{\mu\lambda}\delta A^{\lambda} + A^{\lambda}\delta R_{\mu\lambda}\big) - 2\zeta_R\big(R\delta A_{\mu} + A_{\mu}\delta R\big)\\
&\qquad + \Big(\zeta_S\delta\big({}_{\mathrm{A}}S^{\lambda}_{\mu}\big) + \zeta_F\delta\big({}_{\mathrm{A}}F^{\lambda}{}_{\mu}\big) + \zeta_Q\delta\big({}_{\mathrm{A}}S^{\rho}_{\rho}\big)\delta^{\lambda}_{\mu}\Big)\partial_{\lambda}\ln|\epsilon'_{\mathcal{L}_{\mathrm{A}}}| \\
&\quad + \Big(\zeta_S\,{}_{\mathrm{A}}S^{\lambda}_{\mu} + \zeta_F\,{}_{\mathrm{A}}F^{\lambda}{}_{\mu} + \zeta_Q\,{}_{\mathrm{A}}S^{\rho}_{\rho}\delta^{\lambda}_{\mu}\Big)\partial_{\lambda}\left(\frac{\epsilon''_{\mathcal{L}_{\mathrm{A}}\mathcal{L}_{\mathrm{A}}}}{\epsilon'_{\mathcal{L}_{\mathrm{A}}}}\,\delta\mathcal{L}_{\mathrm{A}} + \frac{\epsilon''_{\mathcal{L}_{\mathrm{A}}\mathcal{L}_{\mathrm{B}}}}{\epsilon'_{\mathcal{L}_{\mathrm{A}}}}\,\delta\mathcal{L}_{\mathrm{B}}\right) = 0\,,
\end{aligned} \tag{6.32a}$$

$$\begin{aligned}
&\bar{\zeta}\delta\big({}_{\mathrm{B}}F^{\lambda}{}_{\mu;\lambda}\big) + \zeta\delta\big({}_{\mathrm{B}}S^{\lambda}_{\lambda;\mu}\big) + 2\zeta_S\big(R_{\mu\lambda}\delta B^{\lambda} + B^{\lambda}\delta R_{\mu\lambda}\big) - 2\zeta_R\big(R\delta B_{\mu} + B_{\mu}\delta R\big)\\
&\qquad + \Big(\zeta_S\delta\big({}_{\mathrm{B}}S^{\lambda}_{\mu}\big) + \zeta_F\delta\big({}_{\mathrm{B}}F^{\lambda}{}_{\mu}\big) + \zeta_Q\delta\big({}_{\mathrm{B}}S^{\rho}_{\rho}\big)\delta^{\lambda}_{\mu}\Big)\partial_{\lambda}\ln|\epsilon'_{\mathcal{L}_{\mathrm{B}}}| \\
&\quad + \Big(\zeta_S\,{}_{\mathrm{B}}S^{\lambda}_{\mu} + \zeta_F\,{}_{\mathrm{B}}F^{\lambda}{}_{\mu} + \zeta_Q\,{}_{\mathrm{B}}S^{\rho}_{\rho}\delta^{\lambda}_{\mu}\Big)\partial_{\lambda}\left(\frac{\epsilon''_{\mathcal{L}_{\mathrm{A}}\mathcal{L}_{\mathrm{B}}}}{\epsilon'_{\mathcal{L}_{\mathrm{B}}}}\,\delta\mathcal{L}_{\mathrm{A}} + \frac{\epsilon''_{\mathcal{L}_{\mathrm{B}}\mathcal{L}_{\mathrm{B}}}}{\epsilon'_{\mathcal{L}_{\mathrm{B}}}}\,\delta\mathcal{L}_{\mathrm{B}}\right) = 0\,.
\end{aligned} \tag{6.32b}$$

We will see shortly that general linear perturbation of $\rho_{\Lambda}g_{\mu\nu} + T^{\mathrm{vec}}_{\mu\nu}$ around the particular exact solution (6.14) contains only a specific combination of $\delta A_{\mu}(x)$ and $\delta B_{\mu}(x)$ like (6.20), namely

$$\delta C_{\mu}(x) = \frac{1}{C_1}\,\delta A_{\mu}(x) - \frac{1}{C_2}\,\delta B_{\mu}(x)\,. \tag{6.33}$$

To derive an equation on δC_{μ} and show that δC_{μ} does not depend on $\delta g_{\mu\nu}$ as $\delta v_{\mathrm{C}}(\tau)$, given in (6.20), does not depend on $\delta h(\tau)$ (see (6.21)), one needs to divide (6.32a) and (6.32b) by C_1 and C_2, respectively, and then subtract them. Using (6.3) and (6.14), we obtain

$$\begin{aligned}
0 &= \bar{\zeta}\mathcal{F}^{\lambda}{}_{\mu;\lambda} + \zeta\mathcal{S}^{\lambda}_{\lambda;\mu} + 2\zeta_S R^{\lambda}_{\mu}\delta C_{\lambda} - 2\zeta_R R\delta C_{\mu} - \frac{\partial_{\lambda}\tau^{2(x-1)}}{\tau^{2(x-1)}}\Big(\zeta_S\mathcal{S}^{\lambda}_{\mu} + \zeta_F\mathcal{F}^{\lambda}{}_{\mu} + \zeta_Q\mathcal{S}^{\rho}_{\rho}\delta^{\lambda}_{\mu}\Big)\\
&\quad - \Big(\zeta_S s^{\lambda}_{\mu} + \zeta_F f^{\lambda}{}_{\mu} + \zeta_Q s^{\rho}_{\rho}\delta^{\lambda}_{\mu}\Big)\partial_{\lambda}\left(\frac{\big(\zeta_S s^{\mu\nu} + \zeta_Q s^{\lambda}_{\lambda}g^{\mu\nu}\big)\mathcal{S}_{\mu\nu} + \zeta_F f^{\mu\nu}\mathcal{F}_{\mu\nu}}{2C_3\tau^{2(x-1)}}\right)
\end{aligned} \tag{6.34}$$

(derivation of this equation is presented in Appendix G), where

$$\begin{aligned}
\mathcal{S}_{\lambda\rho} &\equiv \nabla_{\lambda}\delta C_{\rho} + \nabla_{\rho}\delta C_{\lambda}\,, & \mathcal{F}_{\lambda\rho} &\equiv \nabla_{\lambda}\delta C_{\rho} - \nabla_{\rho}\delta C_{\lambda}\,,\\
s_{\mu\nu} &\equiv {}_{\mathrm{A}}S_{\mu\nu}/C_1 = {}_{\mathrm{B}}S_{\mu\nu}/C_2\,, & f_{\mu\nu} &\equiv {}_{\mathrm{A}}F_{\mu\nu}/C_1 = {}_{\mathrm{B}}F_{\mu\nu}/C_2\,.
\end{aligned} \tag{6.35}$$

and

$$C_3 \equiv \zeta(x+3y)^2 - 6\zeta_S(x+y)y + 6\zeta_R(1-2y)y\,. \tag{6.36}$$

We see that (6.34) does not contain terms depending on $\delta g_{\mu\nu}$, so that δC_{μ} does not depend on the metric perturbation. We stress that this result is a consequence of (6.3) and the particular background solution (6.14).[(6)]

[(6)]Note, for $\zeta_S = \zeta_F = 0$, $\zeta_Q = 1$ and $\zeta_R = -1/2$ with $x = 1$ and $y = 1/2$, (6.34) reduces to $\nabla_{\mu}\nabla^{\lambda}\delta C_{\lambda} = 0$, and this coincides with the equation (5.12) in [103].

Perturbation of vector energy-momentum tensor

The general linear perturbation of the energy-momentum tensor of the vector fields is given by

$$\begin{aligned}\delta T^{\text{vec}}_{\mu\nu} &= \left(\epsilon - \mathcal{L}_{\text{A}}\epsilon'_{\mathcal{L}_{\text{A}}} - \mathcal{L}_{\text{B}}\epsilon'_{\mathcal{L}_{\text{B}}}\right)\delta g_{\mu\nu} + \mathcal{L}_{\text{A}}\epsilon'_{\mathcal{L}_{\text{A}}}\frac{\delta T^{\text{A}}_{\mu\nu}}{\mathcal{L}_{\text{A}}} + \mathcal{L}_{\text{B}}\epsilon'_{\mathcal{L}_{\text{B}}}\frac{\delta T^{\text{B}}_{\mu\nu}}{\mathcal{L}_{\text{B}}} \\ &\quad - \left(\left(\mathcal{L}^2_{\text{A}}\epsilon''_{\mathcal{L}_{\text{A}}\mathcal{L}_{\text{A}}} + \mathcal{L}_{\text{A}}\mathcal{L}_{\text{B}}\epsilon''_{\mathcal{L}_{\text{A}}\mathcal{L}_{\text{B}}}\right)\frac{\delta\mathcal{L}_{\text{A}}}{\mathcal{L}_{\text{A}}} + \left(\mathcal{L}^2_{\text{B}}\epsilon''_{\mathcal{L}_{\text{B}}\mathcal{L}_{\text{B}}} + \mathcal{L}_{\text{A}}\mathcal{L}_{\text{B}}\epsilon''_{\mathcal{L}_{\text{A}}\mathcal{L}_{\text{B}}}\right)\frac{\delta\mathcal{L}_{\text{B}}}{\mathcal{L}_{\text{B}}}\right) g_{\mu\nu} \\ &\quad + \left(\mathcal{L}_{\text{A}}\epsilon''_{\mathcal{L}_{\text{A}}\mathcal{L}_{\text{A}}}T^{\text{A}}_{\mu\nu} + \mathcal{L}_{\text{A}}\epsilon''_{\mathcal{L}_{\text{A}}\mathcal{L}_{\text{B}}}T^{\text{B}}_{\mu\nu}\right)\frac{\delta\mathcal{L}_{\text{A}}}{\mathcal{L}_{\text{A}}} + \left(\mathcal{L}_{\text{B}}\epsilon''_{\mathcal{L}_{\text{B}}\mathcal{L}_{\text{B}}}T^{\text{B}}_{\mu\nu} + \mathcal{L}_{\text{B}}\epsilon''_{\mathcal{L}_{\text{A}}\mathcal{L}_{\text{B}}}T^{\text{A}}_{\mu\nu}\right)\frac{\delta\mathcal{L}_{\text{B}}}{\mathcal{L}_{\text{B}}},\end{aligned} \tag{6.37}$$

where

$$T^{\text{V}}_{\mu\nu} = T^{S}_{\mu\nu}(V) + T^{F}_{\mu\nu}(V) + T^{Q}_{\mu\nu}(V) + T^{R}_{\mu\nu}(V) + T^{\Delta}_{\mu\nu}(V)\,. \tag{6.38}$$

In the case when the background solution is given by (6.14), one has $\mathcal{L}_{\text{B}}T^{\text{A}}_{\mu\nu} = \mathcal{L}_{\text{A}}T^{\text{B}}_{\mu\nu}$. Then using (6.3), (6.37) becomes

$$\begin{aligned}\delta T^{\text{vec}}_{\mu\nu} &= \epsilon(\mathcal{L}_{\text{A}},\mathcal{L}_{\text{B}})\delta g_{\mu\nu} + \mathcal{L}_{\text{A}}\epsilon'_{\mathcal{L}_{\text{A}}}\left(\frac{\delta T^{\text{A}}_{\mu\nu}}{\mathcal{L}_{\text{A}}} - \frac{\delta T^{\text{B}}_{\mu\nu}}{\mathcal{L}_{\text{B}}}\right) \\ &\quad + \mathcal{L}^{-1}_{\text{A}}\left(\mathcal{L}^2_{\text{A}}\epsilon''_{\mathcal{L}_{\text{A}}\mathcal{L}_{\text{A}}} + \mathcal{L}_{\text{A}}\mathcal{L}_{\text{B}}\epsilon''_{\mathcal{L}_{\text{A}}\mathcal{L}_{\text{B}}}\right)\left(T^{\text{A}}_{\mu\nu} - \mathcal{L}_{\text{A}}g_{\mu\nu}\right)\left(\frac{\delta\mathcal{L}_{\text{A}}}{\mathcal{L}_{\text{A}}} - \frac{\delta\mathcal{L}_{\text{B}}}{\mathcal{L}_{\text{B}}}\right).\end{aligned} \tag{6.39}$$

We show in Appendix G that $\delta T^{\text{A}}_{\mu\nu}/\mathcal{L}_{\text{A}} - \delta T^{\text{B}}_{\mu\nu}/\mathcal{L}_{\text{B}}$ and $\delta\mathcal{L}_{\text{A}}/\mathcal{L}_{\text{A}} - \delta\mathcal{L}_{\text{B}}/\mathcal{L}_{\text{B}}$ are functions of δC_μ only. Thus we conclude that $\rho_\Lambda\delta g_{\mu\nu} + \delta T^{\text{vec}}_{\mu\nu}$ does not depend on $\delta g_{\mu\nu}$. In other words, on the right-hand side of the linearized Einstein equations

$$\delta G_{\mu\nu} = 8\pi G\left(\rho_\Lambda\delta g_{\mu\nu} + \delta T^{\text{vec}}_{\mu\nu} + \delta T^{\text{m}}_{\mu\nu}\right), \tag{6.40}$$

there are no terms depending on $\delta g_{\mu\nu}$. Hence, in particular, the Newton gravity law is valid at small spacetime scales, so that we can identify the gravitational constant G with the Newton constant G_{N} as in [103].

In deriving this result, we have used (6.3) and (6.14) resulting in

$$C_\mu(t) \equiv \frac{1}{C_1}A_\mu(t) - \frac{1}{C_2}B_\mu(t) = 0\,. \tag{6.41}$$

However, we have found by considering the homogeneous perturbations around (6.14) that

$$C_\mu(t) = \widetilde{C}_2\,M_{\text{Planck}}\,(t/t_{\text{Planck}})^{x-3y-1} \tag{6.42}$$

subject to $\widetilde{C}_1 = 0$ (see equations (6.20) and (6.22)). If we take this into account, then we find $\delta T^{\text{vec}}_{\mu\nu}$ depends on $\delta g_{\mu\nu}$. Specifically, one can show that if $\zeta_R \neq 0$, then $\delta T^{\text{vec}}_{\mu\nu}$ acquires at small spacetime scales an additional term:

$$\zeta_R\,\widetilde{C}_2\,(t/t_{\text{Planck}})^{3-3y}\partial^4\delta g \tag{6.43}$$

in a symbolic notation.[7] Assuming $\zeta_R \widetilde{C}_2 \sim 1$, one can neglect this term with respect to terms on the left-hand side of (6.40) in the following scales L:

$$L \gg L_{\text{Planck}} \left(t/t_{\text{Planck}}\right)^{\frac{3}{2}(1-y)} . \tag{6.44}$$

Taking $t = t_0 \equiv 1/H_0 \approx 10^{18}$ sec, one calculates $L \gg 10^{-5}$ m for the dust-dominated universe ($y = 2/3$). A planetary system around the metal-poor star HIP 11952 has been recently observed [104]. This star has two planets with semi major axes roughly equaling 10^{11} and 10^{10} meters. The star age t_{star} is approximately 12.8 Gyr or 10^{16} sec, so we have $L \gg 10^{-6}$ m.

If we set $\zeta_R = 0$, then this additional term at small spacetime scales approximately equals

$$\zeta_S \widetilde{C}_2 \, M_{\text{Planck}}^2 \left(t/t_{\text{Planck}}\right)^{1-3y} \partial^2 \delta g \tag{6.45}$$

and, consequently, since $1 \gg (t_{\text{star}}/t_{\text{Planck}})^{1-3y} \approx 10^{-59}$ for $y = 2/3$, it is allowed to neglect a dependence of $\delta T_{\mu\nu}^{\text{vec}}$ on the metric perturbation $\delta g_{\mu\nu}$ in comparison with terms on the left-hand side of (6.40).

The conclusion is that this vector-tensor model reduces to Newton's law of gravity at small spacetime scales, even though one takes into account the homogeneous corrections to the background solution (6.14) subject to $\widetilde{C}_1 = 0$.

[7] We have omitted $M_{\text{Planck}} \left(t/t_{\text{Planck}}\right)^{2-3y} \partial^3 \delta g$, $M_{\text{Planck}}^2 \left(t/t_{\text{Planck}}\right)^{1-3y} \partial^2 \delta g$, $M_{\text{Planck}}^3 \left(t/t_{\text{Planck}}\right)^{-3y} \partial \delta g$ and $M_{\text{Planck}}^4 \left(t/t_{\text{Planck}}\right)^{-1-3y} \delta g$ there as less significant with respect to (6.43).

Chapter 7

Discussion

In Chapter 5, we have dealt with the four-parameter vector model as a generalization of the one-parameter Dolgov model. We have found that it is possible to dynamically cancel the cosmological constant for a broad range of the constant model parameters, even if its value is large. Afterwards the universe expansion is caused only by ordinary matter with the equation of state given in (5.48). The expansion then becomes decelerated.

The solution corresponding to the compensation of the cosmological constant turns out to be attractor. In other words, the late-time evolution of the universe is insensitive to the initial conditions taken from a large phase space domain of the dynamical variables. This proves the cosmological stability of the evolution of such the universe at the classical level. However, this model has no physical relevance, because it does not respect Newton's law of gravity which is well-established at scales from 10^{-4} m [19] up to the size of solar system, i.e. 10^{12} m.

In order to solve the main cosmological constant problem in our own universe, rather than in some hypothetical world, we have to develop a model which does not contradict known experimental facts, such as Newton's law. With this goal in mind, a pair of the vector fields has been considered in Chapter 6 by following the ideas presented in [101].

Strictly speaking, we have found that the particular exact solution (6.14) is unstable with respect to small homogeneous perturbations, unless we choose the initial values of the dynamical variables in a proper way, i.e. $\widetilde{C}_1 = 0$. This means that for the vector fields to cancel the cosmological constant Λ, one has to find a basin of attraction B, such that the vector fields asymptotically behave as (6.14) for any initial values of the dynamical variables taken from B.

At the present stage, we can only say that B must be a hypersurface in the eleven-dimensional phase space spanned by

$$(v_{\text{A,B}}, \dot{v}_{\text{A,B}}, \chi_{\text{A,B}}, \dot{\chi}_{\text{A,B}}, h, \dot{h}, r_{\text{m}}),$$

with points corresponding to $\widetilde{C}_1 = 0$. Any other initial conditions definitely lead to the classical instability of (6.14).

This situation principally differs from that of the case of a scalar model considered by Dolgov and Kawasaki in [107, 108], which is classically unstable as well and, actually, requires fine-tuning. Indeed, they found an asymptotic solution corresponding to $\phi = \phi_0 = \text{const}$ and $H = 1/2t$, where the scalar field plays the role of the Λ-compensator. However, it is clear from equations (2) and (4) in [108] that ϕ_0 is a solution of $U'(\phi_0) = U(\phi_0) + \rho_\Lambda = 0$. This, generally speaking, implies a particular choice of the potential.

Chapter 8

Concluding remarks

The main goal of this work is to study the dynamical cancellation of the large cosmological constant. This approach implies an introduction of a field Ψ or a set of fields Ψ_a, such that Λ-term in the Einstein equations is dynamically compensated by Ψ or Ψ_a at late times of the evolution of the universe.

Violation of Newton's gravity is a very general feature of these kinds of models. Indeed, to dynamically compensate a cosmological constant with an arbitrary value, we have to introduce a massless field Ψ with a field equation depending linearly on the field and with no derivatives of the field higher than the second order, otherwise there may be the Ostrogradsky instability [105, 106]. Then, to obtain an asymptotically stable solution, we have to non-minimally couple Ψ to gravity. Its energy density has then terms like $\dot{\Psi}^2$, $H^2\Psi^2$, $\dot{H}\Psi^2$ and $H\Psi\dot{\Psi}$ in the homogeneous case. Note that if Ψ is a vector or higher-spin field, then its energy density automatically has terms $H^2\Psi^2$ due to the covariant derivative. Hence, the compensation of Λ by Ψ implies that $H \sim 1/t$ and $\Psi \sim t$. In the weak-field limit, these give terms on the right-hand side of the Einstein equations like $t^2\partial^2\delta g$ and $t\partial\delta g$ which do not respect Newton's gravity. It seems that there is only one way out of this difficulty, if one wants the dynamical compensation of the Λ-term, specifically to construct a field model with a rather nonstandard form like that considered in Chapter 6.

A model with a pair of scalar fields with the ϵ-function satisfying (6.3), where the vector Lagrangian (5.2) replaced by (4.5) can be expected to have the same qualitative behavior. Therefore, henceforth, we will talk about Ψ_a ($a = 1, 2$) without specifying their spin.

The two crucial features of the model with Ψ_a considered in Chapter 6 are that the ϵ-function satisfies (6.3) and that $\mathcal{L}(\Psi_a)$ depends only quadratically on Ψ_a. Indeed, this function is symmetric with respect to Ψ_1 and Ψ_2. Therefore, a ratio of their particular exact solutions, i.e. $\overline{\Psi}_1/\overline{\Psi}_2$ or $\overline{\Psi}_2/\overline{\Psi}_1$, must be a constant, that is fixed by Λ as in (6.16b). In the homogeneous case, one has in $T_{\mu\nu}(\Psi_a)$, for instance, a term like

$$H^2\left(\Psi_1^2\frac{\partial\epsilon}{\partial\mathcal{L}(\Psi_1)} + \Psi_2^2\frac{\partial\epsilon}{\partial\mathcal{L}(\Psi_2)}\right), \tag{8.1}$$

which vanishes when $\Psi_a = \overline{\Psi}_a$ as a consequence of (6.3), because $\overline{\Psi}_a$ is proportional to $\mathcal{L}(\overline{\Psi}_a)$ and a coefficient of the proportionality is independent of $\overline{\Psi}_a$. Hence, one has $T_{\mu\nu}(\overline{\Psi}_a) = \epsilon(\overline{\Psi}_a)g_{\mu\nu}$ (cf. (6.2) for (6.14) with (6.16b)).

This result is general, since it does not rely on a specific form of the ϵ-function. Therefore, it may be that there is an ϵ-function (see footnote on page 48) such that $\overline{\Psi}_1$ and $\overline{\Psi}_2$ are asymptotically stable with respect to the small homogeneous perturbations.

If this is the case, the Einstein equations linearized with respect to the general field perturbations take the standard form as that in general relativity. If we take into account small homogeneous perturbations $\delta\overline{\Psi}_a$ around $\overline{\Psi}_a$, then, generally speaking, it will not be so. However, since $\delta\overline{\Psi}_a$ decrease with time in comparison with $\overline{\Psi}_a$ (otherwise $\overline{\Psi}_a$ would be unstable), one is allowed to neglect this additional terms due to $\delta\overline{\Psi}_a$ for sufficiently large time $t_s \ll t_0$.

A possible experimental indication that there might be such a cosmic pair Ψ_a would be if the gravitational Newton law was violated at early times in the evolution of the universe. The primordial gravitational waves would also behave differently from those predicted by general relativity. These issues require, however, further investigation.

In conclusion, it should be emphasized that the analysis described in this thesis concerns only the classical stability of the cosmological solution uncovered above and the consistency of the model with the Newton gravitational law. Another important issues, for instance, quantum stability of the vacuum in the model against spontaneous particle creation [109, 110, 111, 112], must be investigated as well.

Appendices

Appendix A Backreaction of quantum scalar field on metric

The effective action of the scalar field $\Gamma[g]$ can be written as

$$\exp\left(i\Gamma[g]\right) = \int \mathcal{D}\phi \, \exp\left(-\frac{i}{2}\int d^4x\sqrt{-g}\,\phi(\Box + m^2)\phi\right), \tag{2}$$

where we have omitted the surface term assuming that ϕ goes to zero sufficiently rapidly as $x \to 0$, and $\Box$ is the d'Alembert operator with respect to $g_{\mu\nu}$. Following the effective action recipe outlined in [41], we make the Wick rotation: $t \to -i\tau$. Then the metric becomes Euclidean: $\tilde{g}_{\mu\nu}(\tau, x) \equiv -g_{\mu\nu}(t, x)|_{t=-i\tau}$, i.e. real and positive definite, and the differential operator $\Box + m^2$ turns to $-\Box_{\mathrm{E}} + m^2$ which is real, elliptic and self-adjoint, and, as a consequence, has a complete spectrum of eigenfunctions $\phi_n(x)$ with eigenvalues λ_n [42]. The eigenfunctions ϕ_n are normalized in the usual sense with the covariant measure $dx^4\tilde{g}^{1/2}$.

Expanding $\phi(x)$ as a linear combination of $\phi_n(x)$ with coefficients c_n and taking the measure on the field space as a product of $(\mu/\sqrt{2\pi})dc_n$ with μ being a normalization constant with mass dimension, one obtains

$$\Gamma_{\mathrm{E}}[\tilde{g}] = \frac{1}{2}\ln\det\left(-\mu^{-2}(\Box_{\mathrm{E}} - m^2)\right), \tag{3}$$

where the Euclidean effective action $\Gamma_{\mathrm{E}}[\tilde{g}] \equiv -i\Gamma[g]|_{t=-i\tau}$.

Then, let us introduce a Hermitian operator $\hat{M}$ defined on some Hilbert space spanned by vectors $|\psi\rangle$, such that its determinant coincides with that of $-\Box_{\mathrm{E}} + m^2$. This is the case, when the matrix elements of $\hat{M}$ in the coordinate basis $|x\rangle$ are

$$\langle x|\hat{M}|x'\rangle = \left(\tilde{g}(x)\right)^{1/4}\left(-\Box_{\mathrm{E}} + m^2\right)\left(\tilde{g}(x)\right)^{-1/4}\delta(x - x') \tag{4}$$

(see [41] for more details). Then (3) can be rewritten as

$$\Gamma_{\mathrm{E}}[\tilde{g}] = -\frac{1}{2}\lim_{s\to 0}\left(\zeta'_{\hat{M}}(s) + \zeta_{\hat{M}}(s)\ln(\mu^2)\right), \quad \text{where} \quad \zeta_{\hat{M}}(s) \equiv \mathrm{Tr}\left(\hat{M}^{-s}\right) \tag{5}$$

is the so-called zeta function of the operator $\hat{M}$ [42, 43].

Considering $\tilde{g}_{\mu\nu}(x) = \delta_{\mu\nu} + h_{\mu\nu}(x)$, where $|h_{\mu\nu}(x)| \ll 1$, and applying the heat kernel approach [42], after calculations which are quite analogous to those in [41], but without the assumption that m is small, one obtains

$$\zeta_{\hat{M}}(s) = \int d^4x\sqrt{\tilde{g}}\left(J(s,4) + \frac{s}{6}J(s+1,4)R + \mathrm{O}(h^2)\right), \tag{6}$$

where $\tilde{g}^{1/2} = 1 - \frac{1}{2}\delta_{\mu\nu}h^{\mu\nu}(x)$, $R(x) = \delta_{\mu\nu}\partial^2 h^{\mu\nu}(x) - \partial_\mu\partial_\nu h^{\mu\nu}(x)$ and

$$J_{\alpha,4} \equiv \int \frac{d^4k_{\mathrm{E}}}{(2\pi)^4}\frac{1}{(k_{\mathrm{E}}^2+m^2)^\alpha}. \tag{7}$$

Here, the number 4 indicates the dimension of spacetime.

If we substitute (6) in (5) without evaluating the integral in (7), we find that $J(s,4)$, $J(s+1,4)$ and their first derivatives with respect to s are ultraviolet divergent for $s = 0$. In order to regularize these divergences, one can consider a spacetime of the dimension d instead of 4 for which they are finite – this corresponds to dimensional regularization by 't Hooft and Veltman. After we have done this, we go back to the real time and then vary the Lorentzian effective action $\Gamma[g]$ over the metric $g_{\mu\nu}$, where we eventually obtain

$$\langle\hat{T}_{\mu\nu}\rangle = A(\tilde{\mu},d)\,g_{\mu\nu}(x) - 2B(\tilde{\mu},d)\,G_{\mu\nu}(x) + \mathrm{O}(g^2). \tag{8}$$

By definition

$$A(\tilde{\mu},d) \equiv -\frac{1}{2}\frac{m^4}{(4\pi)^{d/2}}\,\Gamma(-d/2)\left(\frac{m}{\tilde{\mu}}\right)^{d-4}, \tag{9a}$$

$$B(\tilde{\mu},d) \equiv +\frac{1}{12}\frac{m^2}{(4\pi)^{d/2}}\,\Gamma(1-d/2)\left(\frac{m}{\tilde{\mu}}\right)^{d-4}, \tag{9b}$$

cf. [40]. There $\tilde{\mu}$ is the 't Hooft scale which gives the correct dimensions for $A(\tilde{\mu},d)$ and $B(\tilde{\mu},d)$.(1)

Direct calculations show that $\mathrm{O}(g^2)$ term appearing in the effective action $\Gamma[g]$ is composed of R^2 and $R_{\mu\nu}R^{\mu\nu}$, and factors in front of them are ultraviolet divergent as well. Therefore, after regularization they must be renormalized (the following terms $\mathrm{O}(g^n)$, $n \geq 3$ are finite). We note that these terms are a source of the trace anomaly and their explicit expressions can be found in [40, 41].

As mentioned above, in particular, $J(s,4)$ and $J'(s,4)$ tend to infinity when $s \to 0$ if we do not evaluate integrals corresponding to them before we take the limit in accordance with the zeta regularization. If we do that, as should be done, then we obtain

$$A(\mu,4) \equiv -\frac{m^4}{64\pi^2}\left(\ln\left(\frac{\mu^2}{m^2}\right) + \frac{3}{2}\right), \tag{10a}$$

$$B(\mu,4) \equiv -\frac{m^2}{192\pi^2}\left(\ln\left(\frac{\mu^2}{m^2}\right) + 1\right). \tag{10b}$$

(1)Note, that $A(\tilde{\mu},d)$ does depend on μ appearing in (6) as a consequence of a property of dimensional regularization giving zero for integrals of the type $\int d^dk_{\mathrm{E}}\,k_{\mathrm{E}}^{-2\alpha}$. But $B(\tilde{\mu},d)$ is also independent of it on account of $sJ(s+1,4) = 0$ for $s = 0$.

Obviously, one does not need to renormalize both $A(\mu, 4)$ and $B(\mu, 4)$, since they are already finite.

It is straightforward to show that in the renormalization scheme known as $\overline{\text{MS}}$ [45] both $A(\tilde{\mu}, 4)$ and $B(\tilde{\mu}, 4)$ given in (9) are equal to $A(\mu, 4)$ and $B(\mu, 4)$, respectively, if we set $\tilde{\mu}/\mu = \exp(\gamma/2)/\sqrt{4\pi}$, where $\gamma \approx 0.577216$ is the Euler-Mascheroni constant.

In conclusion, note that the one-loop effective action $\Gamma[g]$ contains terms proportional to the Einstein-Hilbert and the curvature-squared.(2)

(2)In principle, if one puts Einstein's cosmological constant to zero as well as the Einstein-Hilbert action, then, after a consideration of the one-loop quantum theory on a manifold with a metric $g_{\mu\nu}(x)$, one arrives at the Einstein-Hilbert action with the cosmological and curvature-squared terms. This observation resulted in the idea of Sakharov's induced gravity [46, 47].

Appendix B One-loop vacuum energy

We have found in Section 3.2 that the term in the effective action $\Gamma[g]$ corresponding to the vacuum energy density of the scalar field is given by

$$-\frac{1}{2}\int d^4x\sqrt{-g}\int\frac{d^4k_{\rm E}}{(2\pi)^4}\ln\left(\frac{k_{\rm E}^2+m_0^2}{\mu^2}\right). \tag{B1}$$

Varying this with respect to the metric field $g_{\mu\nu}$, we find a part of $\langle\hat{T}_{\mu\nu}\rangle$ related to the contribution of the scalar field to the cosmological constant, namely $\rho_{\rm V}g_{\mu\nu}$, where by definition

$$\rho_{\rm V} \equiv \frac{1}{2}\int\frac{d^4k_{\rm E}}{(2\pi)^4}\ln\left(\frac{k_{\rm E}^2+m_0^2}{\mu^2}\right). \tag{B2}$$

Let us show that this integral can be expressed as (3.6) and (3.16).

First, since $\ln x=\lim\limits_{s\to0}(dx^s/ds)$, one has

$$\begin{aligned}\rho_{\rm V} &= -\frac{1}{2}\lim_{s\to0}\frac{\partial}{\partial s}\int\frac{d^4k_{\rm E}}{(2\pi)^4}\frac{\mu^{2s}}{(k_{\rm E}^2+m_0^2)^s} = -\frac{1}{2}\lim_{s\to0}\frac{\partial}{\partial s}\int\frac{d^3\mathbf{k}}{(2\pi)^3}\int_{-\infty}^{+\infty}\frac{dk_{\rm E0}}{2\pi}\frac{\mu^{2s}}{(k_{\rm E0}^2+\omega_{\mathbf{k}}^2)^s}\\ &= -\frac{1}{4\pi^{1/2}}\lim_{s\to0}\frac{\partial}{\partial s}\int\frac{d^3\mathbf{k}}{(2\pi)^3}\mu^{2s}\omega_{\mathbf{k}}^{1-2s}\frac{\Gamma(s-1/2)}{\Gamma(s)} = \frac{1}{2}\int\frac{d^3\mathbf{k}}{(2\pi)^3}\omega_{\mathbf{k}},\end{aligned} \tag{B3}$$

where $\omega_{\mathbf{k}}\equiv\sqrt{\mathbf{k}^2+m_0^2}$. Thus, we have shown that (B2) and (3.6) are equal.

Second, since

$$\frac{1}{2\omega_{\mathbf{k}}} = \int_{-\infty}^{+\infty}dk_0\,\delta(k^2-m_0^2)\,\theta(k_0), \tag{B4}$$

one derives from (B3) that

$$\rho_{\rm V} = \int\frac{d^4k}{(2\pi)^3}\,\omega_{\mathbf{k}}^2\,\delta(k^2-m_0^2)\,\theta(k_0) = \int\frac{d^4k}{(2\pi)^3}\,k_0^2\,\delta(k^2-m_0^2)\,\theta(k_0). \tag{B5}$$

Taking into account

$$\int k_\mu k_\nu f(k^2)\,d^4k = \frac{1}{4}\eta_{\mu\nu}\int k^2 f(k^2)\,d^4k, \tag{B6}$$

where $k^2=k_\mu k^\mu$, one obtains

$$\rho_{\rm V} = \frac{m_0^2}{8}\int\frac{d^4k}{(2\pi)^3}\,\delta(k^2-m_0^2), \tag{B7}$$

and then using the Sokhotsky formula, we find

$$\rho_{\rm V} = \frac{m_0^2}{4}\int\frac{d^4k}{(2\pi)^4}\frac{i}{k^2-m_0^2+i\varepsilon} = \frac{Zm_0^2}{4}\int\frac{d^4k}{(2\pi)^4}\frac{i}{Zk^2-Zm_0^2+i\varepsilon}, \tag{B8}$$

where we have taken into account that the principal value integral

$$\mathcal{P}\int_{-\infty}^{+\infty}\frac{dk_0}{k_0^2-\omega_{\mathbf{k}}^2} \tag{B9}$$

vanishes. Thus, we have that (B2) and (3.16) equal each other.

Appendix C Vector energy-momentum tensor

Let us consider the following action

$$S_{\rm A} = -\int d^4x\sqrt{-g}\left(\frac{1}{4}\zeta_S\, S_{\mu\nu}S^{\mu\nu} + \frac{1}{4}\zeta_F\, F_{\mu\nu}F^{\mu\nu} + \frac{1}{4}\zeta_Q\,(S^{\mu}_{\mu})^2 + \zeta_R\, RA_\mu A^\mu\right) . \tag{C1}$$

Variation of this action with respect to the metric field $g_{\mu\nu}$ gives the energy-momentum of the vector field $T^{\rm A}_{\mu\nu}$ which turns out to be a sum of the following terms

$$T^{\rm A}_{\mu\nu} = T^{S}_{\mu\nu} + T^{F}_{\mu\nu} + T^{Q}_{\mu\nu} + T^{R}_{\mu\nu}\,, \tag{C2}$$

where

$$\begin{aligned}
\zeta_S^{-1}T^S_{\mu\nu} &= \tfrac{1}{4}S_{\lambda\rho}S^{\lambda\rho}g_{\mu\nu} - S^{\lambda}_{(\mu}F_{\nu)\lambda} - \tfrac{1}{2}S^{\lambda}_{\lambda}S_{\mu\nu} + 2A_{(\mu}S^{\lambda}_{\nu);\lambda} - A^{\lambda}S_{\mu\nu;\lambda}\,,\\
\zeta_F^{-1}T^F_{\mu\nu} &= -\tfrac{1}{4}\left(4g^{\lambda\rho}F_{\mu\lambda}F_{\nu\rho} - g_{\mu\nu}F^{\lambda\rho}F_{\lambda\rho}\right)\,,\\
\zeta_Q^{-1}T^Q_{\mu\nu} &= -\tfrac{1}{4}\left((S^{\lambda}_{\lambda})^2 + 4A^{\rho}\nabla_{\rho}S^{\lambda}_{\lambda}\right)g_{\mu\nu} + 2A_{(\mu}\nabla_{\nu)}S^{\lambda}_{\lambda}\,,\\
\zeta_R^{-1}T^R_{\mu\nu} &= RA^2g_{\mu\nu} - 2R_{\mu\nu}A^2 - 2RA_\mu A_\nu + 2L_{\mu\nu}A^2\,,
\end{aligned} \tag{C3}$$

where $L_{\mu\nu}$ has been defined as $\nabla_\mu\nabla_\nu - g_{\mu\nu}\nabla^2$. In the case of FRW metric (5.11) and homogeneous configuration of the vector field (5.12), we find

$$\begin{aligned}
\zeta_S^{-1}T^S_{00} &= -\left(\dot A_0 + 3HA_0\right)^2 + 2A_0\left(\ddot A_0 + 6H\dot A_0\right)\\
&\quad + \tfrac{1}{2a^2}\left(\dot A_m - 2HA_m\right)\left(\dot A_n - 2HA_n\right)\delta^{mn}\,,\\
\zeta_S^{-1}T^S_{0i} &= +2\left(\ddot A_0 + 3H\dot A_0 - 3H^2A_0\right)A_i\,,\\
\zeta_S^{-1}T^S_{ij} &= -\left(\dot A_0^2 + 3H^2A_0^2 + 2\left(\dot HA_0^2 + 2HA_0\dot A_0 - \dot A_0^2\right)\right)g_{ij}\\
&\quad - \tfrac{1}{2a^2}\left(\dot A_m - 2HA_m\right)\left(\dot A_n - 2HA_n\right)\delta^{mn}g_{ij}\\
&\quad + \dot A_{(i}\left(\dot A_{j)} - 2HA_{j)}\right) + 2A_{(i}\left(\ddot A_{j)} - 2\left(\dot H + 2H^2\right)A_{j)}\right)\,,
\end{aligned} \tag{C4}$$

$$\begin{aligned}
\zeta_F^{-1}T^F_{00} &= +\tfrac{1}{2a^2}\dot A_m\dot A_n\delta^{mn}\,,\\
\zeta_F^{-1}T^F_{0i} &= 0\,,\\
\zeta_F^{-1}T^F_{ij} &= -\dot A_i\dot A_j - \tfrac{1}{2a^2}\dot A_m\dot A_n\delta^{mn}g_{ij}\,,
\end{aligned} \tag{C5}$$

$$\begin{aligned}
\zeta_Q^{-1}T^Q_{00} &= -\left(\dot A_0 + 3HA_0\right)^2 + 2A_0\left(\ddot A_0 + 3H\dot A_0 + 3\dot HA_0\right)\,,\\
\zeta_Q^{-1}T^Q_{0i} &= +2\left(\ddot A_0 + 3H\dot A_0 + 3\dot HA_0\right)A_i\,,\\
\zeta_Q^{-1}T^Q_{ij} &= -\left(\left(\dot A_0 + 3HA_0\right)^2 + 2A_0\left(\ddot A_0 + 3H\dot A_0 + 3\dot HA_0\right)\right)g_{ij}
\end{aligned} \tag{C6}$$

and

$$\begin{aligned}
\zeta_R^{-1} T_{00}^R &= 12(\dot{H} + 2H^2) A_0^2 - 6H^2 A^2 - 6H\partial_0 A^2 \,, \\
\zeta_R^{-1} T_{0i}^R &= 12(\dot{H} + 2H^2) A_0 \, A_i \,, \\
\zeta_R^{-1} T_{ij}^R &= -2g_{ij}(2\dot{H} + 3H^2 + \partial_0^2 + 2H\partial_0) A^2 + 12(\dot{H} + 2H^2) A_i A_j \,.
\end{aligned} \tag{C7}$$

According to the definitions of the energy density and pressure, we have

$$\rho_{\mathrm{A}} = \rho_{\mathrm{A}_0} + \rho_{\mathrm{A}_i} \,, \quad P_{\mathrm{A}} = P_{\mathrm{A}_0} + P_{\mathrm{A}_i} \,, \tag{C8}$$

where

$$\begin{aligned}
\rho_{\mathrm{A}_0}(t) &= -\zeta \left((\dot{A}_0 + 3HA_0)^2 - 2A_0(\ddot{A}_0 + 3H\dot{A}_0 + 3\dot{H}A_0) \right) \\
&\quad + 6(2\zeta_R - \zeta_S)(\dot{H}A_0^2 - HA_0\dot{A}_0) + 18\zeta_R H^2 A_0^2 \,,
\end{aligned} \tag{C9}$$

$$\begin{aligned}
\rho_{\mathrm{A}_i}(t) &= +\frac{\zeta_S}{2a^2}(\dot{A}_m - 2HA_m)(\dot{A}_n - 2HA_n)\delta^{mn} + \frac{\zeta_F}{2a^2}\dot{A}_m\dot{A}_n\delta^{mn} \\
&\quad + \frac{6\zeta_R}{a^2} H^2 A_m A_n \delta^{mn} + 6\zeta_R H\partial_0\Big(\frac{1}{a^2}A_m A_n\Big)\delta^{mn} \,,
\end{aligned} \tag{C10}$$

$$\begin{aligned}
P_{\mathrm{A}_0}(t) &= +\zeta \left((\dot{A}_0 + 3HA_0)^2 + 2A_0(\ddot{A}_0 + 3H\dot{A}_0 + 3\dot{H}A_0) \right) \qquad \text{(C11)} \\
&\quad + 2(\zeta_R - \zeta_S)\left((2\dot{H} + 3H^2)A_0^2 + 4HA_0\dot{A}_0 \right) + 2(2\zeta_R - \zeta_S)\left(\dot{A}_0^2 + A_0\ddot{A}_0 \right) \,,
\end{aligned}$$

$$\begin{aligned}
P_{\mathrm{A}_i}(t) &= +\frac{\zeta_S}{2a^2}(\dot{A}_m - 2HA_m)(\dot{A}_n - 2HA_n)\delta^{mn} + \frac{\zeta_F}{2a^2}\dot{A}_m\dot{A}_n\delta^{mn} \\
&\quad - 2\zeta_R\left(2\dot{H} + 3H^2 + \partial_0^2 + 2H\partial_0\right)\Big(\frac{1}{a^2}A_m A_n\Big)\delta^{mn} \,.
\end{aligned} \tag{C12}$$

The non-diagonal elements of the vector energy-momentum tensor are given by

$$T_{0i}^{\mathrm{A}} = 2\left(\zeta\frac{d}{dt}(\dot{A}_0 + 3HA_0) + 3\left((2\zeta_R - \zeta_S)\dot{H} + (4\zeta_R - \zeta_S)H^2 \right) A_0 \right) A_i \tag{C13}$$

and

$$\begin{aligned}
T_{ij}^{\mathrm{A,nd}} &= 2\zeta_S A_{(i}(\ddot{A}_{j)} - H\dot{A}_{j)}) \\
&\quad + (\zeta_S - \zeta_F)\dot{A}_i\dot{A}_j + 4(3\zeta_R - \zeta_S)(\dot{H} + 2H^2)A_i A_j \,, \quad i \neq j \,.
\end{aligned} \tag{C14}$$

Appendix D General linear perturbation of vector field equation and energy-momentum tensor

Let us consider small inhomogeneous perturbations of the vector and metric fields around a given background solution

$$\begin{aligned} A_\mu(t) &\rightarrow A_\mu(x) = A_\mu(t) + \delta A_\mu(x)\,, \\ g_{\mu\nu}(t) &\rightarrow g_{\mu\nu}(x) = g_{\mu\nu}(t) + \delta g_{\mu\nu}(x)\,. \end{aligned} \tag{D1}$$

Perturbation of the metric tensor

The general form of the metric perturbation $\delta g_{\mu\nu}$ can be written down as follows

$$ds^2 = \left(1+2\phi\right)dt^2 + 2aS_i dt dx^i - a^2\big((1-2\psi)\delta_{ij} - h_{ij}\big)dx^i dx^j\,, \tag{D2}$$

where $S^i_{,i} = 0$ and $h^i_i = 0$, $h^i_{j,i} = 0$, so that

$$\delta g_{00} = 2\phi\,, \quad \delta g_{0i} = aS_i\,, \quad \delta g_{ij} = a^2\left(2\psi\delta_{ij} + h_{ij}\right) \tag{D3}$$

(see [10] for more details). The general linear perturbation of the Levi-Civita connection is

$$\delta\Gamma^\lambda_{\mu\nu} = \frac{1}{2}g^{\lambda\rho}\left(\delta g_{\rho\mu,\nu} + \delta g_{\rho\nu,\mu} - \delta g_{\mu\nu,\rho}\right) + g_{\rho\sigma}\Gamma^\sigma_{\mu\nu}\delta g^{\lambda\rho}\,, \tag{D4}$$

where we have used an equality

$$\delta g^{\mu\nu} = -g^{\mu\rho}g^{\nu\lambda}\delta g_{\lambda\rho}. \tag{D5}$$

resulting from the variation of $g^{\mu\lambda}g_{\nu\lambda} = \delta^\mu_\nu$. We will need the general linear perturbation of the Ricci tensor $R_{\mu\nu}$ and scalar R:

$$\delta R_{\mu\nu} = \frac{1}{2}\left(\nabla^\lambda\nabla_\nu\delta g_{\mu\lambda} + \nabla^\lambda\nabla_\mu\delta g_{\nu\lambda} - \nabla^2\delta g_{\mu\nu} - g^{\lambda\rho}\nabla_\mu\nabla_\nu\delta g_{\lambda\rho}\right)\,, \tag{D6}$$

$$\delta R = R_{\mu\nu}\delta g^{\mu\nu} + g^{\mu\nu}\delta R_{\mu\nu}\,, \tag{D7}$$

where the covariant derivative ∇_μ is defined with respect to the background metric.

General linear perturbation of vector field equation

The vector field equation of the general vector perturbation δA_μ is

$$\begin{aligned} \bar\zeta\nabla^2\delta A_\mu + (2\zeta - \bar\zeta)\nabla_\mu\nabla^\lambda\delta A_\lambda &= 2\zeta_R A_\mu\delta R - 2\zeta_S A^\lambda\delta R_{\mu\lambda} + 2\zeta\nabla_\mu\left(g^{\lambda\rho}A_\sigma\delta\Gamma^\sigma_{\lambda\rho}\right) \\ &+\bar\zeta\left(g^{\lambda\rho}\left[F_{\sigma\mu}\delta\Gamma^\sigma_{\lambda\rho} + F_{\lambda\sigma}\delta\Gamma^\sigma_{\mu\rho}\right] - F_{\lambda\mu;\rho}\delta g^{\lambda\rho}\right) \\ &+\zeta\left(g^{\lambda\rho}\left[S_{\sigma\rho}\delta\Gamma^\sigma_{\lambda\mu} + S_{\lambda\sigma}\delta\Gamma^\sigma_{\rho\mu}\right] - S_{\lambda\rho;\mu}\delta g^{\lambda\rho}\right) \\ &-2\zeta_S R_{\mu\lambda}A_\rho\delta g^{\lambda\rho} + (\bar\zeta - 2\zeta_S)R^\lambda_\mu\delta A_\lambda + 2\zeta_R R\delta A_\mu. \end{aligned} \tag{D8}$$

General linear perturbation of vector energy-momentum tensor

The perturbation of $T^{\mathrm{A}}_{\mu\nu}$ is given by

$$\delta T^{\mathrm{A}}_{\mu\nu} = \delta T^{S}_{\mu\nu} + \delta T^{F}_{\mu\nu} + \delta T^{Q}_{\mu\nu} + \delta T^{R}_{\mu\nu}\,, \tag{D9}$$

where

$$\begin{aligned}
\zeta_S^{-1}\delta T^{S}_{\mu\nu} &= \frac{1}{4}S_{\lambda\rho}S^{\lambda\rho}\delta g_{\mu\nu} + \frac{1}{2}S^{\lambda\rho}\delta S_{\lambda\rho}g_{\mu\nu} \\
&+\Big(\frac{1}{2}S^{\sigma}_{\lambda}S_{\rho\sigma}g_{\mu\nu} - S_{\lambda(\mu}F_{\nu)\rho} - \frac{1}{2}S_{\mu\nu}S_{\lambda\rho} + 2A_{(\mu}S_{\nu)\lambda;\rho} - A_{\lambda}S_{\mu\nu;\rho}\Big)\delta g^{\lambda\rho} \\
&-\Big(\delta S_{\lambda(\mu}F_{\nu)\rho} + S_{\lambda(\mu}\delta F_{\nu)\rho} + \frac{1}{2}\big[\delta S_{\mu\nu}S_{\lambda\rho} + S_{\mu\nu}\delta S_{\lambda\rho}\big] \\
&-2\big[\delta A_{(\mu}S_{\nu)\lambda;\rho} + A_{(\mu}\delta\big(S_{\nu)\lambda;\rho}\big)\big] + \delta A_{\lambda}S_{\mu\nu;\rho} + A_{\lambda}\delta\big(S_{\mu\nu;\rho}\big)\Big)g^{\lambda\rho}\,,
\end{aligned} \tag{D10}$$

$$\begin{aligned}
\zeta_F^{-1}\delta T^{F}_{\mu\nu} &= \frac{1}{4}F^{\lambda\rho}F_{\lambda\rho}\delta g_{\mu\nu} + \Big(\frac{1}{2}F_{\sigma\lambda}F^{\sigma}{}_{\rho} - F_{\mu\lambda}F_{\nu\rho}\Big)\delta g^{\lambda\rho} + \frac{1}{2}F^{\lambda\rho}\delta F_{\lambda\rho}g_{\mu\nu} \\
&-(\delta F_{\mu\lambda}F_{\nu\rho} + F_{\mu\lambda}\delta F_{\nu\rho})g^{\lambda\rho}\,,
\end{aligned} \tag{D11}$$

$$\begin{aligned}
\zeta_Q^{-1}\delta T^{Q}_{\mu\nu} &= -\frac{1}{4}\left(\left(S^{\lambda}_{\lambda}\right)^2 + 4A^{\rho}S^{\lambda}_{\lambda;\rho}\right)\delta g_{\mu\nu} + 2\delta A_{(\mu}\nabla_{\nu)}S^{\lambda}_{\lambda} \\
&+\left(2A_{(\mu}\nabla_{\nu)}S_{\lambda\rho} - \frac{1}{2}g_{\mu\nu}\left(S^{\sigma}_{\sigma}S_{\lambda\rho} + 2A_{\lambda}S^{\sigma}_{\sigma;\rho} + 2A^{\sigma}S_{\lambda\rho;\sigma}\right)\right)\delta g^{\lambda\rho} \\
&+\Big(2A_{(\mu}\delta\big(\nabla_{\nu)}S_{\lambda\rho}\big) - \frac{1}{2}g_{\mu\nu}\left(S^{\sigma}_{\sigma}\delta S_{\lambda\rho} + 2\delta A_{\lambda}S^{\sigma}_{\sigma;\rho} + 2A^{\sigma}\delta S_{\lambda\rho;\sigma}\right)\Big)g^{\lambda\rho}\,,
\end{aligned} \tag{D12}$$

$$\begin{aligned}
\zeta_R^{-1}\delta T^{R}_{\mu\nu} &= RA^2\delta g_{\mu\nu} + (A^2 g_{\mu\nu} - 2A_{\mu}A_{\nu})\delta R - 2A^2\delta R_{\mu\nu} - 2R(A_{\mu}\delta A_{\nu} + A_{\nu}\delta A_{\mu}) \\
&+(Rg_{\mu\nu} - 2R_{\mu\nu} + 2L_{\mu\nu})(2A^{\lambda}\delta A_{\lambda} + A_{\lambda}A_{\rho}\delta g^{\lambda\rho}) \\
&-2(A^2)_{,\sigma}\left(\delta\Gamma^{\sigma}_{\mu\nu} - g_{\mu\nu}g^{\lambda\rho}\delta\Gamma^{\sigma}_{\lambda\rho}\right) - 2(A^2)_{;\lambda\rho}\delta\left(g_{\mu\nu}g^{\lambda\rho}\right).
\end{aligned} \tag{D13}$$

Newtonian limit

The equation (D8) becomes

$$\bar{\zeta}\partial^2\delta A_{\mu} + (2\zeta - \bar{\zeta})\partial_{\mu}\partial^{\lambda}\delta A_{\lambda} = \delta J_{\mu} \tag{D14}$$

in the Newtonian limit, where by definition

$$\delta J_{\mu} \equiv 2\zeta_R A_{\mu}\delta R - 2\zeta_S A^{\lambda}\delta R_{\mu\lambda} + 2\zeta A_{\sigma}\eta^{\lambda\rho}\delta\Gamma^{\sigma}_{\lambda\rho,\mu}\,. \tag{D15}$$

Its Fourier transform is

$$\delta\widetilde{J}_\mu = 2\zeta_R A_\mu \delta\widetilde{R} - 2\zeta_S A^\lambda \delta\widetilde{R}_{\mu\lambda} + 2i\zeta A_\sigma k_\mu \eta^{\lambda\rho}\delta\widetilde{\Gamma}^\sigma_{\lambda\rho}\,, \tag{D16}$$

where

$$\begin{aligned}
\delta\widetilde{\Gamma}^\lambda_{\mu\nu} &= \tfrac{i}{2}\eta^{\lambda\rho}\big(k_\mu\delta\widetilde{g}_{\nu\rho} + k_\nu\delta\widetilde{g}_{\mu\rho} - k_\rho\delta\widetilde{g}_{\mu\nu}\big)\,,\\
\delta\widetilde{R}_{\mu\nu} &= ik_\lambda\delta\widetilde{\Gamma}^\lambda_{\mu\nu} - ik_\nu\delta\widetilde{\Gamma}^\lambda_{\mu\lambda}\,,\\
\delta\widetilde{R} &= ik_\lambda\eta^{\mu\nu}\delta\widetilde{\Gamma}^\lambda_{\mu\nu} - ik^\mu\delta\widetilde{\Gamma}^\nu_{\mu\nu}\,.
\end{aligned} \tag{D17}$$

Substituting these in $\delta\widetilde{J}_\mu$, one has

$$\begin{aligned}
\delta\widetilde{J}_\mu &= 2i\zeta_R A_\mu\Big[k_\sigma\eta^{\lambda\rho}\delta\widetilde{\Gamma}^\sigma_{\lambda\rho} - k^\lambda\delta\widetilde{\Gamma}^\sigma_{\lambda\sigma}\Big] - 2i\zeta_S A^\kappa\Big[k_\sigma\delta\widetilde{\Gamma}^\sigma_{\mu\kappa} - k_\kappa\delta\widetilde{\Gamma}^\sigma_{\mu\sigma}\Big] + 2i\zeta A_\sigma k_\mu\eta^{\lambda\rho}\delta\widetilde{\Gamma}^\sigma_{\lambda\rho}\\
&= 2i\bigg(\zeta_R A_\mu\Big[k_\sigma\eta^{\lambda\rho} - k^\lambda\delta^\rho_\sigma\Big] - \zeta_S A^\kappa\delta^\lambda_\mu\Big[k_\sigma\delta^\rho_\kappa - k_\kappa\delta^\rho_\sigma\Big] + \zeta A_\sigma k_\mu\eta^{\lambda\rho}\bigg)\delta\widetilde{\Gamma}^\sigma_{\lambda\rho}\,.
\end{aligned} \tag{D18}$$

In a case when $\zeta \neq 0$ and $\bar{\zeta} \neq 0$, we have

$$\delta\widetilde{A}_\mu = -\frac{1}{\bar{\zeta}k^2}\left(\delta^\nu_\mu + \frac{\bar{\zeta}-2\zeta}{2\zeta}\frac{k_\mu k^\nu}{k^2}\right)\delta\widetilde{J}_\nu(k) \equiv 2i\Sigma^{(\lambda\rho)}_{\mu\sigma}\delta\widetilde{\Gamma}^\sigma_{\lambda\rho}\,, \tag{D19}$$

where $\Sigma^{\lambda\rho}_{\mu\sigma}$ has been defined as

$$\begin{aligned}
\Sigma^{\lambda\rho}_{\mu\sigma} &\equiv -\frac{1}{\bar{\zeta}k^2}\left(\delta^\nu_\mu + \frac{\bar{\zeta}-2\zeta}{2\zeta}\frac{k_\mu k^\nu}{k^2}\right)\\
&\times\bigg(\zeta_R A_\nu\Big[k_\sigma\eta^{\lambda\rho} - k^\lambda\delta^\rho_\sigma\Big] - \zeta_S A^\kappa\delta^\lambda_\nu\Big[k_\sigma\delta^\rho_\kappa - k_\kappa\delta^\rho_\sigma\Big] + \zeta A_\sigma k_\nu\eta^{\lambda\rho}\bigg)\,.
\end{aligned} \tag{D20}$$

The Fourier transform of the energy-momentum tensor is given by

$$\begin{aligned}
\delta\widetilde{T}_{\mu\nu}(k) &= \zeta_S\bigg(-A_\mu\Big[k^2\delta\widetilde{A}_\nu + k_\nu k^\lambda\delta\widetilde{A}_\lambda + 2iA_\sigma k^\lambda\delta\widetilde{\Gamma}^\sigma_{\nu\lambda}\Big] - A_\nu\Big[k^2\delta\widetilde{A}_\mu + k_\mu k^\lambda\delta\widetilde{A}_\lambda\\
&\quad + 2iA_\sigma k^\lambda\delta\widetilde{\Gamma}^\sigma_{\mu\lambda}\Big] + A_\kappa k^\kappa\Big[k_\mu\delta\widetilde{A}_\nu + k_\nu\delta\widetilde{A}_\mu + 2iA_\sigma\delta\widetilde{\Gamma}^\sigma_{\mu\nu}\Big]\bigg)\\
&\quad - 2\zeta_Q\left(A_\mu k_\nu + A_\nu k_\mu - \eta_{\mu\nu}A^\kappa k_\kappa\right)\left(k^\lambda\delta\widetilde{A}_\lambda + iA_\sigma\eta^{\lambda\rho}\delta\widetilde{\Gamma}^\sigma_{\lambda\rho}\right)\\
&\quad + \zeta_R\bigg(\Big[A^2\eta_{\mu\nu} - 2A_\mu A_\nu\Big]\delta\widetilde{R} - 2A^2\delta\widetilde{R}_{\mu\nu} + 2l_{\mu\nu}\Big[2A^\lambda\delta\widetilde{A}_\lambda + A^\lambda A^\rho\delta\widetilde{g}_{\lambda\rho}\Big]\bigg)\,,
\end{aligned} \tag{D21}$$

where by definition $l_{\mu\nu} \equiv k^2\eta_{\mu\nu} - k_\mu k_\nu$. Substituting $\delta\widetilde{A}_\mu$ expressed via $\delta\widetilde{\Gamma}^\mu_{\lambda\rho}$ in $\delta\widetilde{T}_{\mu\nu}$, one finds

$$\begin{aligned}
\delta\widetilde{T}_{\mu\nu}(k) &= \left(\zeta_S M^{\lambda\rho}_{\mu\nu\sigma} + \zeta_Q N^{\lambda\rho}_{\mu\nu\sigma} + \zeta_R K^{\lambda\rho}_{\mu\nu\sigma}\right)\eta^{\sigma\kappa}\left(k_\kappa\delta\widetilde{g}_{\lambda\rho} - k_\lambda\delta\widetilde{g}_{\rho\kappa} - \kappa_\rho\delta\widetilde{g}_{\lambda\kappa}\right)\\
&\quad + 2\zeta_R\left(k^2\eta_{\mu\nu} - k_\mu k_\nu\right)A^\lambda A^\rho\delta\widetilde{g}_{\lambda\rho}\,,
\end{aligned} \tag{D22}$$

where

$$M_{\mu\nu\sigma}^{\lambda\rho} \equiv A_\kappa k^\kappa\Big(k_\mu\Sigma_{\nu\sigma}^{(\lambda\rho)} + k_\nu\Sigma_{\mu\sigma}^{(\lambda\rho)} - A_\sigma\delta_\mu^{(\lambda}\delta_\nu^{\rho)}\Big) - A_\mu\Big(k^2\Sigma_{\nu\sigma}^{(\lambda\rho)} + k_\nu k^\kappa\Sigma_{\kappa\sigma}^{(\lambda\rho)} + A_\sigma k^{(\lambda}\delta_\nu^{\rho)}\Big)$$
$$-A_\nu\Big(k^2\Sigma_{\mu\sigma}^{(\lambda\rho)} + k_\mu k^\kappa\Sigma_{\kappa\sigma}^{(\lambda\rho)} + A_\sigma k^{(\lambda}\delta_\mu^{\rho)}\Big), \quad \text{(D23a)}$$

$$N_{\mu\nu\sigma}^{\lambda\rho} \equiv -\left(A_\mu k_\nu + A_\nu k_\mu - \eta_{\mu\nu}A_\kappa k^\kappa\right)\left(2k^\alpha\Sigma_{\alpha\sigma}^{(\lambda\rho)} + A_\sigma\eta^{\lambda\rho}\right), \quad \text{(D23b)}$$

$$K_{\mu\nu\sigma}^{\lambda\rho} \equiv \frac{1}{2}\left(A^2\eta_{\mu\nu} - 2A_\mu A_\nu\right)\left(k_\sigma\eta^{\lambda\rho} - k^{(\lambda}\delta_\sigma^{\rho)}\right) - A^2\left(k_\sigma\delta_\mu^{(\lambda}\delta_\nu^{\rho)} - k_\nu\delta_\mu^{(\lambda}\delta_\sigma^{\rho)}\right)$$
$$+4\left(k^2\eta_{\mu\nu} - k_\mu k_\nu\right)A^\kappa\Sigma_{\kappa\sigma}^{(\lambda\rho)}. \quad \text{(D23c)}$$

It can also be rewritten as

$$\delta\widetilde{T}_{\mu\nu}(k) = \Big(k^\sigma\Big[\zeta_S M_{\mu\nu\sigma}^{\lambda\rho} + \zeta_Q N_{\mu\nu\sigma}^{\lambda\rho} + \zeta_R K_{\mu\nu\sigma}^{\lambda\rho}\Big] + 2\zeta_R\Big[k^2\eta_{\mu\nu} - k_\mu k_\nu\Big]A^\lambda A^\rho\Big)\delta\widetilde{g}_{\lambda\rho}$$
$$-2\kappa_\rho\eta^{\sigma\kappa}\Big(\zeta_S M_{\mu\nu\sigma}^{\lambda\rho} + \zeta_Q N_{\mu\nu\sigma}^{\lambda\rho} + \zeta_R K_{\mu\nu\sigma}^{\lambda\rho}\Big)\delta\widetilde{g}_{\lambda\kappa}, \quad \text{(D24)}$$

where we have taken into account that $M_{\mu\nu\sigma}^{\lambda\rho}$, $N_{\mu\nu\sigma}^{\lambda\rho}$ and $K_{\mu\nu\sigma}^{\lambda\rho}$ are symmetric with respect to λ and ρ.

Appendix E Energy-momentum tensor of vector fields

The variation of the action (6.1) with respect to the metric $g_{\mu\nu}$ is found to be

$$\delta_g S[g,A,B] = -\int d^4x\sqrt{-g}\left(-\frac{1}{2}\epsilon\, g_{\mu\nu}\delta g^{\mu\nu} + \sum_{\rm V}\frac{\partial\epsilon}{\partial\mathcal{L}_{\rm V}}\delta_g\mathcal{L}_{\rm V}\right) \tag{E1}$$

$$= \int d^4x\sqrt{-g}\left(\frac{1}{2}\left(\epsilon - \sum_{\rm V}\frac{\partial\epsilon}{\partial\mathcal{L}_{\rm V}}\mathcal{L}_{\rm V}\right)g_{\mu\nu} + \sum_{\rm V}\frac{\partial\epsilon}{\partial\mathcal{L}_{\rm V}}\left(\frac{1}{2}\mathcal{L}_{\rm V}g_{\mu\nu} - \frac{\delta\mathcal{L}_{\rm V}}{\delta g^{\mu\nu}}\right)\right)\delta g^{\mu\nu},$$

Consequently, the energy-momentum tensor of the vector fields is

$$T^{\rm vec}_{\mu\nu} = \Big(\epsilon - \sum_{\rm V}\mathcal{L}_{\rm V}\,\epsilon'_{\mathcal{L}_{\rm V}}\Big)g_{\mu\nu} + \sum_{\rm V}\epsilon'_{\mathcal{L}_{\rm V}}\left(T^S_{\mu\nu}(V) + T^F_{\mu\nu}(V) + T^Q_{\mu\nu}(V) + T^R_{\mu\nu}(V) + T^\Delta_{\mu\nu}(V)\right), \tag{E2}$$

where

$$T^\Delta_{\mu\nu}(V) = \Big(\zeta_S(2S^\lambda_{(\mu}V_{\nu)} - S_{\mu\nu}V^\lambda) + \zeta_Q S^\rho_\rho\left(2V_{(\mu}\delta^\lambda_{\nu)} - g_{\mu\nu}V^\lambda\right)\Big)\partial_\lambda\ln|\epsilon'_{\mathcal{L}_{\rm V}}| + 2\zeta_R\left(\epsilon'_{\mathcal{L}_{\rm V}}\right)^{-1}\left(L_{\mu\nu}\left(\epsilon'_{\mathcal{L}_{\rm V}}V^2\right) - \epsilon'_{\mathcal{L}_{\rm V}}L_{\mu\nu}(V^2)\right). \tag{E3}$$

All terms in the energy-momentum tensor (E2) in the case of the flat Friedmann-Robertson-Walker metric and the homogeneous configuration of the vector field are known from Appendix C, except $T^\Delta_{\mu\nu}$:

$$T^\Delta_{00}(V) = 2\left(\zeta(\dot V_0 + 3HV_0)V_0 - 3\zeta_S HV_0^2 - 3\zeta_R HV^2\right)\partial_0\ln|\epsilon'_{\mathcal{L}_{\rm V}}|, \tag{E4a}$$

$$T^\Delta_{0i}(V) = 2\left(\zeta(\dot V_0 + 3HV_0) - 3\zeta_S HV_0\right)V_i\partial_0\ln|\epsilon'_{\mathcal{L}_{\rm V}}|, \tag{E4b}$$

$$T^\Delta_{ij}(V) = 2\zeta_S V_{(i}(\dot V_{j)} - 2HV_{j)})\partial_0\ln|\epsilon'_{\mathcal{L}_{\rm V}}| - 2g_{ij}\Big(\zeta_Q(\dot V_0 + 3HV_0)V_0 + \zeta_S HV_0^2 + \zeta_R V^2\left(\partial_0 + \partial_0\ln|\epsilon'_{\mathcal{L}_{\rm V}}|\right) + 2\zeta_R\partial_0 V^2 + 2\zeta_R HV^2\Big)\partial_0\ln|\epsilon'_{\mathcal{L}_{\rm V}}|. \tag{E4c}$$

Appendix F Expansion coefficients

In Section 6.2, we have defined π-coefficients which are given here:

π^a-coefficients

$$\pi_1^a = a\alpha(1+3y)\Big(x\alpha+3y\big[\alpha+3\beta+2\alpha\beta\big]\Big)C_1^4+3b\Big(\alpha^2\big[1+3y\big]\big[x+3y\big]$$
$$+\alpha\beta\big[21+10\alpha+y(51+42\alpha)\big]y+12\beta^2\big[3+2\alpha\big]^2y^2\Big)C_2^4, \tag{F1a}$$

$$\pi_2^a = \frac{1}{4}\Big(4a\alpha\big[1+3y\big]\Big[\alpha(1-x+3y)(x+3y)+3\beta(\alpha+6[2+\alpha]y)y\Big]C_1^4$$
$$+b\Big[\alpha(x+3y)(2\alpha-8\alpha x^2-9+3[3+2\alpha][1-2y]x+6[3+2\alpha(4+9y)]y)$$
$$+3\beta(27[2x+y-2][1-2y]+3\alpha[3+4y(23+66y)]$$
$$+2\alpha^2[7+18y(5+14y)])y+144\beta^2(3+2\alpha)(\alpha+6[2+\alpha]y)y^2\Big]C_2^4\Big), \tag{F1b}$$

$$\pi_3^a = \frac{1}{16}\Big(a\big[1+3\big]\Big[\alpha(x+3y)(3[1-x][1-2y]+2\alpha[4x^2-3x-1-12(1+x)y])$$
$$-9\beta(6x[1-2y]+[2+\alpha-y][6(1+4\alpha)y+2\alpha-3])y\Big]C_1^4+b\Big[16\alpha^2x^4+4\alpha$$
$$\times(6[1+\alpha]y-3-4\alpha)x^3-\alpha(8\alpha-9+9[4+14\alpha+(30\alpha-13$$
$$+6[3+4\alpha]y)y]y)x+54\beta(1+[5-14y]y)xy+x^2(\alpha[3+8\alpha-3(5+6\alpha$$
$$-6[1-8\alpha]y)y]-216\beta[1-2y]y)-3y(\alpha+3\beta(2-y+\alpha))(8\alpha-9$$
$$+3[15+50\alpha+16\alpha\beta+6(20\alpha-3+8\beta(7+4\alpha))y]y)\Big]C_2^4\Big). \tag{F1c}$$

π^b-coefficients

$$\pi_1^b = \pi_1^a+2\Big(aC_1^4-bC_2^2\Big) \tag{F2a}$$
$$\times\Big(\alpha^2\big[1+3y\big]\big[x+3y\big]+3\alpha\beta\big[9+4\alpha+3(7+6\alpha)y\big]y+18\beta^2\big[3+2\alpha\big]^2y^2\Big),$$

$$\pi_2^b = \pi_2^a-\frac{1}{4}\Big(aC_1^4-bC_2^4\Big)$$
$$\times\Big(\alpha\big[x+3y\big]\Big[9(1-x)(1-2y)+2\alpha(1-(5-4x)x-12[1+3y]y)\Big]$$
$$-3\beta\Big[27(2x+y-2)(1-2y)+3\alpha(3+4[19+54y]y)$$

$$+\, 2\alpha^2\big(5+72[1+3y]y\big)\Big]y - 144\beta^2\big[3+2\alpha\big]\big[\alpha+6(2+\alpha)y\big]y^2\Big)\,, \tag{F2b}$$

$$\begin{aligned}\pi_3^b \;=\;& \pi_3^a + \frac{1}{8}\Big(aC_1^4 - bC_2^4\Big)\Big(6\alpha\big[2y-1-2\alpha\big]x^3 + 8\alpha^2x^4 - 3\alpha\Big[\alpha - 1 + \big(5+13\alpha \\ &+ 12[3y-2+\alpha]y\big)y\Big]x + 54\beta\big[1+(3-10y)y\big]xy + x^2\Big[\alpha\big(3+7\alpha-6 \\ &\times [1+12\alpha y]y\big) - 108\beta\big(1-2y\big)y\Big] - 9\big[\alpha+3\beta(2-y+\alpha)\big]\Big[\alpha - 1 \\ &+ 4\big(2+5\alpha+2\alpha\beta+3[4\alpha-1+14\beta+8\alpha\beta]y\big)y\Big]y\Big)\,. \end{aligned} \tag{F2c}$$

π^c**-coefficients**

$$\pi_1^c \;=\; -2\Big(\alpha x + 3\big[\alpha+3\beta+2\alpha\beta\big]y\Big)^2\Big(aC_1^4 + bC_2^4\Big)\,, \tag{F3a}$$

$$\begin{aligned}\pi_2^c \;=\;& \frac{1}{8}\Big(\alpha\big[x+3y\big]\Big[9\big(1-x\big)\big(1-2y\big) + 2\alpha\big(4x^2 - x - 3 - 12[3-x]y - 72y^2\big)\Big] \\ &- 27\beta\Big[3\big(2x+y-2\big)\big(1-2y\big) + \alpha\big(1+4[9+26y]y\big) + 2\alpha^2\big(1+4y\big) \\ &\times \big(1+8y\big)\Big]y - 144\beta^2\big[3+2\alpha\big]\Big[\alpha+6\big(2+\alpha\big)y\Big]y^2\Big)\Big(aC_1^4 + bC_2^4\Big)\,, \end{aligned} \tag{F3b}$$

$$\begin{aligned}\pi_3^c \;=\;& -\frac{1}{16}\Big(\alpha\big[x+3y\big]\Big[3\big(1-x\big)\big(1-2y\big)\big(2+2x-3y\big) + \alpha\big(8x^3 - 4x^2 + x - 5 \\ &- 6[15+x]y - 72[3+x]y^2\big)\Big] + 9\beta\Big[3\big(2+2x-3y\big)\big(2-2x-y\big)\big(1-2y\big) \\ &- 4\alpha\big(1+[49+3(41-18y)y]y\big) - \alpha^2\big(1+4y\big)\big(5+78y\big)\Big]y \\ &- 216\beta^2\big[2-y+\alpha\big]\Big[\alpha+3\big(7+4\alpha\big)y\Big]y^2\Big)\Big(aC_1^4 + bC_2^4\Big)\,. \end{aligned} \tag{F3c}$$

π^h**-coefficients**

$$\pi_1^h \;=\; -3\beta\big(3+\alpha\big)\Big(\alpha x + 3\big[\alpha+3\beta+2\alpha\beta\big]y\Big)\Big(aC_1^4 - bC_2^4\Big)\,, \tag{F4a}$$

$$\begin{aligned}\pi_2^h \;=\;& -\frac{3}{8}\Big(\alpha\big[x+3y\big]\Big[7\big(1-x\big) + 2\alpha\big(5-x+12y\big)\Big] + \beta\Big[8\alpha\big(9+3x \\ &+ 3\alpha + 2\alpha x\big)x - 9\big(7-10\alpha[7+4\alpha]\big)y^2 + 3\big(42-42x+\alpha[177+62\alpha]\big)y\Big] \\ &+ 24\beta^2\big[3+2\alpha\big]\Big[4\big(3+\alpha\big) + 3\big(7+3\alpha\big)y\Big]y\Big)\Big(aC_1^4 - bC_2^4\Big)\,, \end{aligned} \tag{F4b}$$

$$
\begin{aligned}
\pi_3^h \;=\; & -\frac{3}{16}\Big(\alpha\,[x+3y]\Big[(1-x)(19+18y)+2\alpha(11+[1-4x]x+24y)\Big] \\
& +\beta\Big[9(2-2x-y)(19+18y)y+3\alpha(32x^2+3[95+4(43+12y)y]y) \\
& +2\alpha^2(32x^2+3[49+120y]y)\Big]+48\beta^2\,[3+2\alpha]\Big[2(3+\alpha)+3(8+3y \\
& +3\alpha)y\Big]y\Big)\Big(aC_1^4-bC_2^4\Big)\,. \qquad \text{(F4c)}
\end{aligned}
$$

π^r-coefficients

$$
\begin{aligned}
\pi_1^r \;=\; & \Pi_r\Big(\alpha\,[x+3y]\,[2\alpha(x-12y-5)-3(1-x)]-3\beta\,[18-18x+\alpha(93+38\alpha) \\
& -3(3-78\alpha-56\alpha^2)y]\,y-72\beta^2\,[3+2\alpha]\,[9+5\alpha]\,y^2\Big)\,, \qquad \text{(F5a)}
\end{aligned}
$$

$$
\begin{aligned}
\pi_2^r \;=\; & \frac{\Pi_r}{2}\Big(\alpha\,[x+3y]\Big[3(1-x)(6+2x-13y)+\alpha(19-3[5-4x]x+48y)\Big] \\
& -3\beta\Big[9(6+2x-13y)(2x+y-2)-6\alpha(12-y(13+168y)) \\
& -\alpha^2(25-432y^2)\Big]+144\beta^2\Big[9+\alpha(3-\alpha)-3(27+26\alpha+6\alpha^2)y\Big]\Big)\,, \qquad \text{(F5b)}
\end{aligned}
$$

$$
\begin{aligned}
\pi_3^r \;=\; & \frac{\Pi_r}{4}\Big(6\alpha\,[1-2\alpha]\,x^4+6\Big[\alpha(1+2\alpha)(1-y)+18\beta y\Big]x^3-\Big[\alpha(13\alpha-3 \\
& +3[12+25\alpha+3(11+4\alpha)y]y)-54\beta(2-7y)y\Big]x^2-\Big[3\alpha(5-[29 \\
& -9(5+3y)y]y)-\alpha^2(13+3[80+3(71+60y)y]y)-54\beta(1-[16+6\beta \\
& +(13-12\beta)y]y)y\Big]x-3\Big[9\beta(10-[57+12\beta-(8+9y-6\beta[219+86y]) \\
& \times y]y)-\alpha^2(13+39\beta+3[93+23\beta(13+4\beta)+288(1+2\beta)^2y^2+12(27 \\
& +\beta[109+100\beta])y]y)-3\alpha(11\beta-5+[26+36\beta^2(37+84\beta)y^2 \\
& +\beta(581+186\beta)+3(3+8\beta[89+177\beta])y]y)\Big]y\Big)\,. \qquad \text{(F5c)}
\end{aligned}
$$

where by definition

$$
\Pi_r \;\equiv\; \frac{aC_1^4-bC_2^4}{4(1+3y)^2\,(\alpha x+3[\alpha+3\beta+2\alpha\beta]y)^2\,C_1^2C_2^2}\,. \qquad \text{(F6)}
$$

π^χ-coefficients

$$\pi_1^\chi = 2 - 2x + 3y\,, \tag{F7a}$$

$$\pi_2^\chi = \frac{y}{\alpha\bar{\zeta}}\Big(\alpha\bar{\zeta}\big[1 - 2x + 2y\big] - 2\big[2\beta\zeta(2\alpha + 3)x + 9y - 5\alpha - 15\big]\Big)\,. \tag{F7b}$$

Appendix G General linear perturbation of vector field equations and energy-momentum tensor of vector fields

General linear perturbation of vector field equations

The linearized vector field equations (6.5a) and (6.5b) are given in (6.32). Let us show that if we divide (6.32a) and (6.32b) by C_1 and C_2, respectively, and then subtract them, we obtain (6.34) by using (6.3) and (6.14).

Indeed, if we do that, we obtain a quite complicated equation which we call here as the big-equation. This equation is rather huge, therefore we do not write it down explicitly. Let us consider all of its terms separately. Begin with

$$\frac{1}{C_1}\delta\big({}_{\mathrm{A}}S^{\lambda}_{\lambda;\mu}\big) - \frac{1}{C_2}\delta\big({}_{\mathrm{B}}S^{\lambda}_{\lambda;\mu}\big) \;=\; \nabla_\mu\left(\left[\nabla_\lambda\delta C_\rho + \nabla_\rho\delta C_\lambda - 2C_\sigma\delta\Gamma^{\sigma}_{\lambda\rho}\right]g^{\lambda\rho} + {}_{\mathrm{C}}S_{\lambda\rho}\delta g^{\lambda\rho}\right),$$

where by definition

$$C_\mu(t) \;\equiv\; \frac{1}{C_1}A_\mu(t) - \frac{1}{C_2}B_\mu(t)\,, \tag{G1}$$

cf. (6.20). It immediately follows from (6.14) that $C_\mu(t) = 0$. Thus we obtain

$$\frac{1}{C_1}\delta\big({}_{\mathrm{A}}S^{\lambda}_{\lambda;\mu}\big) - \frac{1}{C_2}\delta\big({}_{\mathrm{B}}S^{\lambda}_{\lambda;\mu}\big) \;=\; \nabla_\mu\big(\mathcal{S}_{\lambda\rho}g^{\lambda\rho}\big)\,, \quad \text{where} \tag{G2}$$

$$\mathcal{S}_{\lambda\rho} \;\equiv\; \nabla_\lambda\delta C_\rho + \nabla_\rho\delta C_\lambda\,, \tag{G3}$$

Analogously, one finds

$$\frac{1}{C_1}\delta\big({}_{\mathrm{A}}F^{\mu}{}_{\mu;\lambda}\big) - \frac{1}{C_2}\delta\big({}_{\mathrm{B}}F^{\mu}{}_{\mu;\lambda}\big) \;=\; \nabla^\lambda\mathcal{F}_{\lambda\mu}\,, \quad \text{where} \tag{G4}$$

$$\mathcal{F}_{\lambda\mu} \;\equiv\; \nabla_\lambda\delta C_\mu - \nabla_\mu\delta C_\lambda\,. \tag{G5}$$

We have omitted ${}_{\mathrm{C}}F_{\lambda\mu;\rho}\delta g^{\lambda\rho}$ and $-g^{\lambda\rho}\big({}_{\mathrm{C}}F_{\sigma\mu}\delta\Gamma^{\sigma}_{\lambda\rho} + {}_{\mathrm{C}}F_{\lambda\sigma}\delta\Gamma^{\sigma}_{\mu\rho}\big)$ there, since (G1) is zero. Obviously, terms $C^\lambda\delta R_{\mu\lambda}$ and $C_\mu\delta R$ in the big-equation vanish as well. The next term we are moving on is

$$\frac{\zeta_S}{C_1}\,\delta\big({}_{\mathrm{A}}S^{\lambda}_{\mu}\big)\partial_\lambda\ln\big|\,\epsilon'_{\mathcal{L}_{\mathrm{A}}}\big| - \frac{\zeta_S}{C_2}\,\delta\big({}_{\mathrm{B}}S^{\lambda}_{\mu}\big)\partial_\lambda\ln\big|\,\epsilon'_{\mathcal{L}_{\mathrm{B}}}\big|\,. \tag{G6}$$

Let us treat one of the derivative of the logarithms in detail, namely

$$\partial_\lambda\ln\big|\,\epsilon'_{\mathcal{L}_{\mathrm{A}}}\big| \;=\; \frac{\mathcal{L}_{\mathrm{A}}\epsilon''_{\mathcal{L}_{\mathrm{A}}\mathcal{L}_{\mathrm{A}}}}{\epsilon'_{\mathcal{L}_{\mathrm{A}}}}\frac{\partial_\lambda\mathcal{L}_{\mathrm{A}}}{\mathcal{L}_{\mathrm{A}}} + \frac{\mathcal{L}_{\mathrm{B}}\epsilon''_{\mathcal{L}_{\mathrm{A}}\mathcal{L}_{\mathrm{B}}}}{\epsilon'_{\mathcal{L}_{\mathrm{A}}}}\frac{\partial_\lambda\mathcal{L}_{\mathrm{B}}}{\mathcal{L}_{\mathrm{B}}}\,. \tag{G7}$$

Substituting the background solution (6.14) in $\mathcal{L}_{\mathrm{A}}$ and $\mathcal{L}_{\mathrm{B}}$, one has

$$\mathcal{L}_{\mathrm{A,B}} \;=\; C_3\,C_{1,2}^2\,\tau^{2(x-1)} \quad\Rightarrow\quad \frac{\partial_\lambda\mathcal{L}_{A,B}}{\mathcal{L}_{\mathrm{A,B}}} \;=\; \frac{\partial_\lambda\tau^{2(x-1)}}{\tau^{2(x-1)}}\,, \tag{G8}$$

where C_3 has been defined as

$$C_3 \equiv \zeta(x+3y)^2 - 6\zeta_S(x+y)y + 6\zeta_R(1-2y)y\,, \tag{G9}$$

so that it depends only on the ζ-coefficients and $w_{\rm m}$. Thus one has

$$\partial_\lambda \ln|\,\epsilon'_{\mathcal{L}_{\rm A}}| = \left(\frac{\mathcal{L}_A\epsilon''_{\mathcal{L}_{\rm A}\mathcal{L}_{\rm A}}}{\epsilon'_{\mathcal{L}_{\rm A}}} + \frac{\mathcal{L}_{\rm B}\epsilon''_{\mathcal{L}_{\rm A}\mathcal{L}_{\rm B}}}{\epsilon'_{\mathcal{L}_{\rm A}}}\right)\frac{\partial_\lambda\tau^{2(x-1)}}{\tau^{2(x-1)}} \tag{G10}$$

and, consequently,

$$\frac{\zeta_S}{C_1}\,\delta(\,{}_{\rm A}S^\lambda_\mu)\partial_\lambda\ln|\,\epsilon'_{\mathcal{L}_{\rm A}}| - \frac{\zeta_S}{C_2}\,\delta(\,{}_{\rm B}S^\lambda_\mu)\partial_\lambda\ln|\,\epsilon'_{\mathcal{L}_{\rm B}}| \tag{G11}$$

$$= \frac{\zeta_S}{C_1}\frac{\partial_\lambda\tau^{2(x-1)}}{\tau^{2(x-1)}}\left(\Delta_{\rm A}+\Delta_{\rm B}\right)\delta(\,{}_{\rm A}S^\lambda_\mu) + \zeta_S\frac{\partial_\lambda\tau^{2(x-1)}}{\tau^{2(x-1)}}\left(\frac{\mathcal{L}_{\rm A}\epsilon''_{\mathcal{L}_{\rm A}\mathcal{L}_{\rm B}} + \mathcal{L}_{\rm B}\epsilon''_{\mathcal{L}_{\rm B}\mathcal{L}_{\rm B}}}{\epsilon'_{\mathcal{L}_{\rm B}}}\right)\left(g^{\lambda\nu}\mathcal{S}_{\nu\mu}\right),$$

where for making formulas more transparent, we have defined $\Delta_{\rm A}$ and $\Delta_{\rm B}$ as

$$\Delta_{\rm A} \equiv \frac{\mathcal{L}_{\rm A}\epsilon''_{\mathcal{L}_{\rm A}\mathcal{L}_{\rm A}}}{\epsilon'_{\mathcal{L}_{\rm A}}} - \frac{\mathcal{L}_{\rm A}\epsilon''_{\mathcal{L}_{\rm A}\mathcal{L}_{\rm B}}}{\epsilon'_{\mathcal{L}_{\rm B}}}\,, \quad \Delta_{\rm B} \equiv \frac{\mathcal{L}_{\rm B}\epsilon''_{\mathcal{L}_{\rm A}\mathcal{L}_{\rm B}}}{\epsilon'_{\mathcal{L}_{\rm A}}} - \frac{\mathcal{L}_{\rm B}\epsilon''_{\mathcal{L}_{\rm B}\mathcal{L}_{\rm B}}}{\epsilon'_{\mathcal{L}_{\rm B}}}\,. \tag{G12}$$

We have omitted ${}_{\rm C}S_{\mu\nu}\delta g^{\nu\lambda} - 2g^{\lambda\nu}C_\rho\delta\Gamma^\rho_{\mu\nu}$ in the last round brackets, because $C_\mu(t)=0$ in a case of our particular solution. The same sequence of steps gives similar results of the rest of terms, i.e. $\delta(\,{}_{\rm V}F^\lambda{}_\mu)$ and $\delta^\lambda_\mu\delta(\,{}_{\rm V}S^\rho_\rho)$, multiplied by $\partial_\lambda\ln(\epsilon'_{\mathcal{L}_{\rm V}})$, but instead of $\delta(\,{}_{\rm V}S^\lambda_\mu)$ and $g^{\lambda\nu}\mathcal{S}_{\nu\mu}$, there must be $\delta(\,{}_{\rm V}F^\lambda{}_\mu)$, $g^{\lambda\nu}\mathcal{F}_{\nu\mu}$ and $\delta^\lambda_\mu\delta(\,{}_{\rm V}S^\rho_\rho)$, $\delta^\lambda_\mu g^{\nu\rho}\mathcal{S}_{\nu\rho}$, respectively. Now we are left with one of the terms like

$$\frac{\zeta_S}{C_1}\,{}_{\rm A}S^\lambda_\mu\partial_\lambda\left(\frac{\epsilon''_{\mathcal{L}_{\rm A}\mathcal{L}_{\rm A}}}{\epsilon'_{\mathcal{L}_{\rm A}}}\delta\mathcal{L}_{\rm A} + \frac{\epsilon''_{\mathcal{L}_{\rm A}\mathcal{L}_{\rm B}}}{\epsilon'_{\mathcal{L}_{\rm A}}}\delta\mathcal{L}_{\rm B}\right) - \frac{\zeta_S}{C_2}\,{}_{\rm B}S^\lambda_\mu\partial_\lambda\left(\frac{\epsilon''_{\mathcal{L}_{\rm A}\mathcal{L}_{\rm B}}}{\epsilon'_{\mathcal{L}_{\rm B}}}\delta\mathcal{L}_{\rm A} + \frac{\epsilon''_{\mathcal{L}_{\rm B}\mathcal{L}_{\rm B}}}{\epsilon'_{\mathcal{L}_{\rm B}}}\delta\mathcal{L}_{\rm B}\right) \tag{G13}$$

in the big-equation. Since $C_\mu(t)=0$, one has $C_2\,{}_{\rm A}S^\lambda_\mu = C_1\,{}_{\rm B}S^\lambda_\mu$, therefore one can rewrite the above expression as follows

$$\frac{\zeta_S}{C_1}\,{}_{\rm A}S^\lambda_\mu\partial_\lambda\left(\Delta_{\rm A}\frac{\delta\mathcal{L}_{\rm A}}{\mathcal{L}_{\rm A}} + \Delta_{\rm B}\frac{\delta\mathcal{L}_{\rm B}}{\mathcal{L}_{\rm B}}\right). \tag{G14}$$

Note, that $\Delta_{\rm A}$ and $\Delta_{\rm B}$ are constants. Indeed, for the background solution (6.14), one has $\epsilon'_{\mathcal{L}_{\rm V}} \sim \tau^{2(1-x)}$ and $\epsilon''_{\mathcal{L}_{\rm V}\mathcal{L}_{\rm V}} \sim \tau^{4(1-x)}$, so that $\mathcal{L}_{\rm V}\epsilon''_{\mathcal{L}_{\rm V}\mathcal{L}_{\rm V}}/\epsilon'_{\mathcal{L}_{\rm V}} \sim 1$, consequently, $\Delta_{\rm A,B}$ are constants. Hence, one can write

$$\partial_\lambda\left(\Delta_{\rm A}\frac{\delta\mathcal{L}_{\rm A}}{\mathcal{L}_{\rm A}}\right) = \Delta_{\rm A}\partial_\lambda\left(\frac{\delta\mathcal{L}_{\rm A}}{\mathcal{L}_{\rm A}}\right) \tag{G15}$$

and taking into account

$$\frac{\delta\mathcal{L}_{\rm A}}{\mathcal{L}_{\rm A}} - \frac{\delta\mathcal{L}_{\rm B}}{\mathcal{L}_{\rm B}} = \frac{\tau^{2(1-x)}}{C_3}\left(P^{\mu\nu}_S\mathcal{S}_{\mu\nu} + P^{\mu\nu}_F\mathcal{F}_{\mu\nu} + 2\zeta_R R C^\mu\delta C_\mu\right), \tag{G16}$$

where $P^{\mu\nu}_S$ and $P^{\mu\nu}_F$ have been defined as

$$P^{\mu\nu}_S \equiv \frac{\zeta_S}{2}\,s^{\mu\nu} + \frac{\zeta_Q}{2}\,s^\lambda_\lambda g^{\mu\nu}\,, \quad P^{\mu\nu}_F \equiv \frac{\zeta_F}{2}\,f^{\mu\nu}\,, \tag{G17}$$

where

$$s_{\mu\nu} \equiv {}_{\rm A}S_{\mu\nu}/C_1 = {}_{\rm B}S_{\mu\nu}/C_2\,, \quad f_{\mu\nu} \equiv {}_{\rm A}F_{\mu\nu}/C_1 = {}_{\rm B}F_{\mu\nu}/C_2\,, \tag{G18}$$

we obtain

$$\begin{aligned}
&\frac{\zeta_S}{C_1}\,{}_{\rm A}S^{\lambda}_{\mu}\partial_\lambda\left(\frac{\epsilon''_{\mathcal{L}_A\mathcal{L}_A}}{\epsilon'_{\mathcal{L}_A}}\delta\mathcal{L}_A + \frac{\epsilon''_{\mathcal{L}_A\mathcal{L}_B}}{\epsilon'_{\mathcal{L}_A}}\delta\mathcal{L}_B\right) - \frac{\zeta_S}{C_2}\,{}_{\rm B}S^{\lambda}_{\mu}\partial_\lambda\left(\frac{\epsilon''_{\mathcal{L}_A\mathcal{L}_B}}{\epsilon'_{\mathcal{L}_B}}\delta\mathcal{L}_A + \frac{\epsilon''_{\mathcal{L}_B\mathcal{L}_B}}{\epsilon'_{\mathcal{L}_B}}\delta\mathcal{L}_B\right)\\
&= \frac{\zeta_S\,{}_{\rm A}S^{\lambda}_{\mu}}{C_1}\left(\Delta_A+\Delta_B\right)\partial_\lambda\left(\frac{\delta\mathcal{L}_A}{\mathcal{L}_A}\right) - \frac{\zeta_S\,{}_{\rm A}S^{\lambda}_{\mu}}{C_1}\left(\frac{\mathcal{L}_B\epsilon''_{\mathcal{L}_A\mathcal{L}_B}}{\epsilon'_{\mathcal{L}_A}} - \frac{\mathcal{L}_B\epsilon''_{\mathcal{L}_B\mathcal{L}_B}}{\epsilon'_{\mathcal{L}_B}}\right)\\
&\times\partial_\lambda\left(\frac{P^{\mu\nu}_S\mathcal{S}_{\mu\nu} + P^{\mu\nu}_F\mathcal{F}_{\mu\nu} + 2\zeta_R R C^\mu\delta C_\mu}{C_3\tau^{2(x-1)}}\right)\,.
\end{aligned} \tag{G19}$$

Making replacement $\zeta_S\,{}_{\rm A}S^{\lambda}_{\mu}$ by $\zeta_F\,{}_{\rm A}F^{\lambda}{}_{\mu}$ and $\zeta_Q\,{}_{\rm A}S^{\rho}_{\rho}\delta^{\lambda}_{\mu}$, we find the last two terms in the big-equation. Now let us calculate a sum $\Delta_A + \Delta_B$. From the definitions of the summands (G12) and the properties of the ϵ-function (6.3), we have

$$\Delta_A + \Delta_B = \frac{\mathcal{L}_A^2\epsilon''_{\mathcal{L}_A\mathcal{L}_A} + \mathcal{L}_A\mathcal{L}_B\epsilon''_{\mathcal{L}_A\mathcal{L}_B}}{\mathcal{L}_A\epsilon'_{\mathcal{L}_A}} - \frac{\mathcal{L}_A\mathcal{L}_B\epsilon''_{\mathcal{L}_A\mathcal{L}_B} + \mathcal{L}_B^2\epsilon''_{\mathcal{L}_B\mathcal{L}_B}}{\mathcal{L}_B\epsilon'_{\mathcal{L}_B}} = 0\,. \tag{G20}$$

Thus, using (G20), the big-equation becomes

$$\begin{aligned}
&\bar{\zeta}\mathcal{F}^{\lambda}{}_{\mu;\lambda} + \zeta\mathcal{S}^{\lambda}_{\lambda;\mu} + 2\zeta_S R^{\lambda}_{\mu}\delta C_\lambda - 2\zeta_R R\delta C_\mu - \frac{\partial_\lambda\tau^{2(x-1)}}{\tau^{2(x-1)}}\left(\zeta_S\mathcal{S}^{\lambda}_{\mu} + \zeta_F\mathcal{F}^{\lambda}{}_{\mu} + \zeta_Q\mathcal{S}^{\rho}_{\rho}\delta^{\lambda}_{\mu}\right)\\
&\qquad - \left(\zeta_S s^{\lambda}_{\mu} + \zeta_F f^{\lambda}{}_{\mu} + \zeta_Q s^{\rho}_{\rho}\delta^{\lambda}_{\mu}\right)\partial_\lambda\left(\frac{P^{\mu\nu}_S\mathcal{S}_{\mu\nu} + P^{\mu\nu}_F\mathcal{F}_{\mu\nu}}{C_3\tau^{2(x-1)}}\right) = 0
\end{aligned} \tag{G21}$$

what was to be proved.

General linear perturbation of energy-momentum tensor

We have found in Chapter 6 that

$$\begin{aligned}
\delta T^{\rm vec}_{\mu\nu} &= \epsilon(\mathcal{L}_A,\mathcal{L}_B)\delta g_{\mu\nu} + \mathcal{L}_A\epsilon'_{\mathcal{L}_A}\left(\frac{\delta T^{A}_{\mu\nu}}{\mathcal{L}_A} - \frac{\delta T^{B}_{\mu\nu}}{\mathcal{L}_B}\right)\\
&+\mathcal{L}_A^{-1}\left(\mathcal{L}_A^2\epsilon''_{\mathcal{L}_A\mathcal{L}_A} + \mathcal{L}_A\mathcal{L}_B\epsilon''_{\mathcal{L}_A\mathcal{L}_B}\right)\left(T^{A}_{\mu\nu} - \mathcal{L}_A g_{\mu\nu}\right)\left(\frac{\delta\mathcal{L}_A}{\mathcal{L}_A} - \frac{\delta\mathcal{L}_B}{\mathcal{L}_B}\right)\,,
\end{aligned} \tag{G22}$$

where $\delta\mathcal{L}_A/\mathcal{L}_A - \delta\mathcal{L}_B/\mathcal{L}_B$ is given in (G16), so it does not depend on the metric perturbation, and

$$\frac{\delta T^{A}_{\mu\nu}}{\mathcal{L}_A} - \frac{\delta T^{B}_{\mu\nu}}{\mathcal{L}_B} = C_3^{-1}\tau^{2(1-x)}\sum_f \zeta_f^{-1}\left(\frac{1}{C_1^2}\,\delta T^f_{\mu\nu}(A) - \frac{1}{C_2^2}\,\delta T^f_{\mu\nu}(B)\right)\,. \tag{G23}$$

Here now f runs over S, F, Q, R, Δ, where by definition $\zeta_\Delta \equiv 1$. Terms in (G23) can be found by use Appendices B and C:

$$\frac{\delta T^S_{\mu\nu}(A)}{C_1^2} - \frac{\delta T^S_{\mu\nu}(B)}{C_2^2} = \frac{1}{2} s^{\lambda\rho} \mathcal{S}_{\lambda\rho} g_{\mu\nu} - \Big(\mathcal{S}_{\lambda(\mu} f_{\nu)\rho} + s_{\lambda(\mu} \mathcal{F}_{\nu)\rho} + \frac{1}{2}\big[\mathcal{S}_{\mu\nu} s_{\lambda\rho} + s_{\mu\nu}\mathcal{S}_{\lambda\rho}\big]$$

$$-2\big[\delta C_{(\mu} s_{\nu)\lambda;\rho} + a_{(\mu}\mathcal{S}_{\nu)\lambda;\rho}\big] + \delta C_\lambda s_{\mu\nu;\rho} + a_\lambda \mathcal{S}_{\mu\nu;\rho}\Big) g^{\lambda\rho}, \qquad \text{(G24a)}$$

$$\frac{\delta T^F_{\mu\nu}(A)}{C_1^2} - \frac{\delta T^F_{\mu\nu}(B)}{C_2^2} = \frac{1}{2} f^{\lambda\rho}\mathcal{F}_{\lambda\rho} g_{\mu\nu} - f_\nu{}^\lambda \mathcal{F}_{\mu\lambda} - f_\mu{}^\lambda \mathcal{F}_{\nu\lambda}, \qquad \text{(G24b)}$$

$$\frac{\delta T^Q_{\mu\nu}(A)}{C_1^2} - \frac{\delta T^Q_{\mu\nu}(B)}{C_2^2} = \Big(\delta C_\mu \nabla_\nu s_{\lambda\rho} + \delta C_\nu \nabla_\mu s_{\lambda\rho} + a_\mu \mathcal{S}_{\lambda\rho;\nu} + a_\nu \mathcal{S}_{\lambda\rho;\mu}$$

$$-\frac{1}{2} g_{\mu\nu}\big[s^\sigma_\sigma \mathcal{S}_{\lambda\rho} + 2\delta C_\lambda s^\sigma_{\sigma;\rho} + 2a^\sigma \mathcal{S}_{\sigma\lambda;\rho}\big]\Big) g^{\lambda\rho}, \qquad \text{(G24c)}$$

$$\frac{\delta T^R_{\mu\nu}(A)}{C_1^2} - \frac{\delta T^R_{\mu\nu}(B)}{C_2^2} = -4R a_{(\mu}\delta C_{\nu)} + 2\big(R g_{\mu\nu} - 2R_{\mu\nu} + 2L_{\mu\nu}\big)\big(a^\lambda \delta C_\lambda\big), \qquad \text{(G24d)}$$

$$\frac{\delta T^\Delta_{\mu\nu}(A)}{C_1^2} - \frac{\delta T^\Delta_{\mu\nu}(B)}{C_2^2} = -2(x-1)\frac{\partial^\lambda \tau}{\tau}\Big(\zeta_S\big[2\mathcal{S}_{\lambda(\mu} a_{\nu)} + 2s_{\lambda(\mu}\delta C_{\nu)} - a_\lambda \mathcal{S}_{\mu\nu} - s_{\mu\nu}\delta C_\lambda\big]$$

$$+\ \zeta_Q \mathcal{S}^\sigma_\sigma\big[2a_{(\mu} g_{\nu)\lambda} - g_{\mu\nu} a_\lambda\big] + \zeta_Q s^\sigma_\sigma\big[2\delta C_{(\mu} g_{\nu)\lambda} - g_{\mu\nu}\delta C_\lambda\big]\Big)$$

$$+\ \Big(\zeta_S\big[2s^\lambda_{(\mu} a_{\nu)} - s_{\mu\nu} a^\lambda\big] + \zeta_Q s^\rho_\rho\big[2a_{(\mu}\delta^\lambda_{\nu)} - g_{\mu\nu} a^\lambda\big]\Big)\partial_\lambda\left(\Delta_A \frac{\delta\mathcal{L}_A}{\mathcal{L}_A} + \Delta_B \frac{\delta \mathcal{L}_B}{\mathcal{L}_B}\right)$$

$$-\ 4\zeta_R L_{\mu\nu}(a^\lambda \delta C_\lambda) + 4\zeta_R \tau^{2(x-1)} L_{\mu\nu}(\tau^{2(1-x)} a^\lambda \delta C_\lambda) - 2\zeta_R \tau^{2(x-1)} L_{\mu\nu}(\tau^2) \qquad \text{(G24e)}$$

$$\times\ \left(\Delta_A \frac{\delta\mathcal{L}_A}{\mathcal{L}_A} + \Delta_B \frac{\delta \mathcal{L}_B}{\mathcal{L}_B}\right) + 2\zeta_R \tau^{2(x-1)} L_{\mu\nu}\left(\tau^{2(1-x)} a^2 \left(\Delta_A \frac{\delta\mathcal{L}_A}{\mathcal{L}_A} + \Delta_B \frac{\delta \mathcal{L}_B}{\mathcal{L}_B}\right)\right).$$

Using the fact that $\Delta_A + \Delta_B$ is zero and (G16), one has

$$\Delta_A \frac{\delta\mathcal{L}_A}{\mathcal{L}_A} + \Delta_B \frac{\delta \mathcal{L}_B}{\mathcal{L}_B} = \Delta_A\left(\frac{\delta\mathcal{L}_A}{\mathcal{L}_A} - \frac{\delta \mathcal{L}_B}{\mathcal{L}_B}\right)$$

$$= \Delta_A \frac{\tau^{2(1-x)}}{C_3}\Big(P_S^{\mu\nu}\mathcal{S}_{\mu\nu} + P_F^{\mu\nu}\mathcal{F}_{\mu\nu} + 2\zeta_R R C^\mu \delta C_\mu\Big), \qquad \text{(G25)}$$

i.e. (G23) does not depend on the metric perturbation $\delta g_{\mu\nu}$.

Acknowledgements

I am indebted to my supervisor, Prof. Dr. F.R. Klinkhamer, for introducing me to this scientific problem and the time he spent on my person.

During my study, I was surrounded by very responsive people: Mikhail Rogal, Marco Schreck, Markus Schwarz, Joel Weller as well as Renate Weiss. I am also sincerely thankful to them.

Special thanks goes to my wife, Natalia, for constructive criticism of my person and her unending patience. I am grateful to Rimma for the cheerful start of days.

My father, Anatoly Emelyanov, and my mother, Marina Emelyanova, are the two persons which made this possible at all. Their direct and indirect contributions to my education are invaluable.

References

[1] (a) A.G. Riess *et al.*, "Observational evidence from Supernovae for an accelerating universe and a cosmological constant," The Astronomical Journal **116**, 1009–1038 (1998); arXiv:astro-ph/9805201. (b) S. Perlmutter *et al.*, "Measurements of Ω and Λ from 42 high-redshift Supernovae," The Astrophysical Journal **517**, 565–586 (1999); arXiv:astro-ph/9812133.

[2] N. Jarosik *et al.*, "Seven-year Wilkinson Microwave Anisotropy Probe (WMAP) observations: sky maps, systematics errors, and basic results," The Astrophysical Journal Supplement Series **192** (2011); arXiv:astro-ph/1001.4758.

[3] W. Percival *et al.*, "Baryon acoustic oscillations in the Sloan digital sky survey data release 7 galaxy sample," Monthly Notices of the Royal Astronomical Society **401**, 2148–2168 (2010); arXiv:astro-ph/0907.1660.

[4] F.R. Klinkhamer and G.E. Volovik, "Self-tuning vacuum variable and cosmological constant," Phys. Rev. D **77**, 085015 (2008); arXiv:gr-qc/0711.3170.

[5] A.D. Dolgov, (a) "Field model with a dynamic cancellation of the cosmological constant," Pis'ma Zh. Eksp. Teor. Fiz. **41**, 6, 280–282 (1985); (b) "Higher spin fields and the problem of the cosmological constant," Phys. Rev. D **55**, 5881–5885 (1997); arXiv:astro-ph/9608175.

[6] F.R. Klinkhamer and G.E. Volovik, "Towards a solution of the cosmological constant problem," JETP Lett. **91**, 259 (2010); arXiv:hep-th/0907.4887.

[7] W.N. Cottingham and D.A. Greenwood, *An introduction to the standard model of particle physics* (Cambridge University Press 2007).

[8] M.E. Peskin and D.V. Schroeder, *An introduction to quantum field theory* (Addison–Wesley Publishing Company 1995).

[9] S.N. Vergeles, *Lectures on theory of gravity* (Moscow Institute of Physics and Technology 2001 (in Russian).

[10] V.F. Mukhanov, *Physical foundations of cosmology* (Cambridge University Press 2005).

[11] L.D. Landau and E.M. Lifshitz, *The Classical theory of fields. Volume 2 of course of theoretical physics* (Fourth Revised English Edition, Butterworth–Heinemann 1975).

[12] J.H. Oort, "The force exerted by the stellar system on the direction perpendicular to the galactic plane and some related problems," Bulletin of the Astronomical Institutes of the Netherlands **6**, 249–287 (1932).

[13] (a) F. Zwicky, "Die Rotverschiebung von extragalaktischen Nebeln," Helvetica Physica Acta **6**, 110–127 (1933); "The redshift of extragalactic nebulae," General Relativity and Gravitation **41**, 1, 207–224 (2009). (b) F. Zwicky, "On the masses of nebulae and of clusters of nebulae," The Astrophysical Journal **86**, 3, 217–246 (1937).

[14] (a) S. Matarrese *et al.*, *Dark matter and dark energy. A challenge for modern cosmology* (Springer 2011). (b) M. Livio, *The dark universe. Matter, energy and gravity* (Cambridge University Press 2003).

[15] M. Milgrom, "A modification of the Newtonian dynamics as a possible alternative to the hidden mass hypothesis," The Astrophysical Journal **270**, 365–370 (1983).

[16] A.H. Guth, "Inflationary universe: A possible solution to the horizon and flatness problems," Phys. Rev. D **23**, 347 (1981).

[17] A.D. Linde, "Chaotic inflation," Phys. Lett. B **129**, 177 (1983).

[18] P.J. Mohr, B.N. Taylor and D.B. Newell, "CODATA recommended values of the fundamental physical constants: 2006," Rev. Mod. Phys. **80** (2008); arXiv:atom-ph/1203.5425.

[19] E.Q. Adelberger, B.R. Heckel and A.E. Nelson, "Tests of the gravitational inverse-square law," Anne. Rev. Nucl. Sci. **53**, 77–121 (2003); arXiv:hep-ph/0307284.

[20] A. Einstein, "Kosmologische Betrachtungen zur allgemeinen Relativitätstheorie," Sitzungsber. Preuss. Akad. Wiss. **1**, 142–152 (1917).

[21] N. Straumann, "On the cosmological constant problem and the astronomical evidence for a homogeneous energy density with negative pressure," arXiv:astro-ph/0203330.

[22] E. Hubble, "A relation between distance and radial velocity among extra-galactic nabulae," Proceedings of the National Academy of Sciences of the United States of America **15**, Issue 3, 168–174 (1929).

[23] (a) P.J.E. Peebles and B. Ratra, "Cosmology with a time-variable cosmological "constant"," The Astrophysical Journal **325**, L17–20 (1988). (b) C. Wetterich, "Cosmology and the fate of dilatation symmetry," Nucl. Phys. B **302**, 668–696 (1988). (c) R.R. Caldwell, R. Dave and P.J. Steinhardt,

"Cosmological imprint of an energy component with general equation of state," Phys. Rev. Lett. **80**, 8, 1582–1585 (1998); arXiv:astro-ph/9708069. (d) P.G. Ferreira and M. Joyce, "Structure formation with a self-tuning scalar field," Phys. Rev. Lett. **79**, 24, 4740–4743 (1997); arXiv:astro-ph/9707286. (e) I. Zlatev, L. Wang and P.J. Steinhardt, "Quintessence, cosmic coincidence, and the cosmological constant," Phys. Rev. Lett. **82**, 5, 896–899 (1999); arXiv:astro-ph/9807002.

[24] C. Armendariz-Picon, V. Mukhanov and P.J. Steinhardt, "Dynamical solution to the problem of a small cosmological constant and late-time cosmic acceleration," Phys. Rev. Lett. **85**, 21, 4438–4441 (2000); arXiv:astro-ph/0004134, "Essentials of k-essence," Phys. Rev. D **63**, 103510 (2001); arXiv:astro-ph/0006373.

[25] T.P. Sotiriou and V. Faraoni, "$f(R)$ theories of gravity," Rev. Mod. Phys. **82** 451–497 (2010); arXiv:gr-qc/0805.1726.

[26] Ya.B. Zel'dovich, "Cosmological constant and elementary particles," JETP letters **6**, 316 (1967); "The cosmological constant and the theory of elementary particles," Soviet Physics Uspekhi **11**(3), 381–393 (1968).

[27] W. Pauli and F. Villars, "On the invariant regularization in relativistic quantum theory," Rev. Mod. Phys. **21**, 434 (1949).

[28] A. Zee, *Quantum field theory in a nutshell* (Princeton University Press 2003).

[29] P.W. Milonni, *The quantum vacuum. An introduction to quantum electrodynamics* (Academic Press, Inc. 1994).

[30] H.A. Bethe, "The electromagnetic shift of energy levels," Phys. Rev. **72**, 4, 339–341 (1947).

[31] (a) W.E. Lamb, Jr. and R.C. Retherford, "Fine structure of the hydrogen atom by a microwave method," Phys. Rev. **72**, 3, 241–243 (1947); (b) W.E. Lamb, Jr. and R.C. Retherford, "Fine structure of the hydrogen atom. Part I," Phys. Rev. **79**, 549–572 (1950); (c) W.E. Lamb, Jr. and R.C. Retherford, "Fine structure of the hydrogen atom. Part II," Phys. Rev. **81**, 222–532 (1951); (d) W.E. Lamb, "Fine structure of the hydrogen atom. Part III," Phys. Rev. **85**, 259–276 (1952); (e)W.E. Lamb, Jr. and R.C. Retherford, "Fine structure of the hydrogen atom. Part IV," Phys. Rev. **86**, 1014–1022 (1952).

[32] J. Schwinger, "On quantum-electrodynamics and the magnetic moment of the electron," Phys. Rev. **73**, 416–417 (1948).

[33] H. Dehmelt, "Experiments with an isolated subatomic particle at rest," Rev. Mod. Phys. **62**, 525–530 (1990).

[34] H.B.G. Casimir, "On the attraction between two perfectly conducting plates," Proc. K. Ned. Akad. Wet. **51**, 793–795 (1948).

[35] S.K. Lamoreaux, "Demonstration of the Casimir force in the 0.6 to 6 μm range," Phys. Rev. Lett. **78**, 5–8 (1997).

[36] U. Mohideen and A. Roy, "Precision measurement of the Casimir force from 0.1 to 0.9 μm," Phys. Rev. Lett. **81**, 4549–4552 (1998).

[37] M. Bordag, U. Mohideen and V.M. Mostepanenko, "New developments in the Casimir effect," Phys. Rep. **353**, 1–205 (2001); arXiv:physics/9805038.

[38] R.L. Jaffe, "Casimir effect and the quantum vacuum," Phys. Rev. D **72**, 021301 (2005); arXiv:hep-th/0503158.

[39] S. Räsänen, "Vacuum energy and dynamical symmetry breaking in curved spacetime," arXiv:gr-qc/1203.6259.

[40] N.D. Birrell and P.C.W. Davies, *Quantum fields in curved space* (Cambridge University Press 1982).

[41] V.F. Mukhanov and S. Winitzki, *Introduction to quantum effects in gravity* (Cambridge University Press 2007).

[42] S.W. Hawking, "Zeta function regularization of path integrals in curved spacetime," Commun. Math. Phys. **55**, 133–148 (1977).

[43] E. Elizalde, S.D. Odintsov, A. Romeo, A.A. Bytsenko and S. Zerbini, *Zeta regularization techiques with applications* (World Scientific Publishing Co. Pte. Ltd. 1994).

[44] G. 't Hooft and M.J. Veltman, "Regularization and renormalization of gauge fields," Nucl. Phys. B **44**, 189 (1972).

[45] J.C. Collins, *Renormalization* (Cambridge University Press 1984).

[46] A.D. Sakharov, "Vacuum quantum fluctuations in curved space and the theory of gravitation," Sov. Phys. Dokl. **12** (1968).

[47] M. Visser, "Sakharov's induced gravity: a modern perspective," Mod. Phys. Lett. A **17**, 977–992 (2002); arXiv:gr-qc/0204062.

[48] P.J.E. Peebles and B. Ratra, "The cosmological constant and dark energy," Rev. Mod. Phys. **75**, 559–606 (2003); arXiv:astro-ph/0207347.

[49] S. Hollands and R.M. Wald, "Quantum field theory is not merely quantum mechanics applied to low energy effective degrees of freedom," General Relativity and Gravitation **36**, 12 (2004); arXiv:gr-qc/0405082.

[50] S.P. Martin, "A supersymmetry primer," arXiv:hep-ph/9709356.

[51] I.J.R. Aitchison, *Supersymmetry in particle physics. An elementary introduction* (Cambridge University Press 2007).

[52] E.Kh. Akhmedov, "Vacuum energy and relativistic invariance," arXiv:hep-th/0204048.

[53] G. Ossola and A. Sirlin, "Considerations concerning the contributions of fundamental particles to the vacuum energy density," Eur. Phys. J. C **31**, 165–175 (2003); arXiv:hep-ph/0305050.

[54] J.F. Koksma and T. Prokopec, "The cosmological constant and Lorentz invariance of the vacuum state," arXiv:gr-qc/1105.6296.

[55] C.M. Will, "The confrontation between general relativity and experiment," Living Rev. Relativity **9**, 3 (2006).

[56] D. Bailin and A. Love, *Introduction to gauge field theory* (IOP Publishing Ltd. 1993).

[57] B. Kastening, "Four-loop vacuum energy β function in $O(N)$ symmetric scalar theory," Phys. Rev. D **54**, 6, 3965–3975 (1996); arXiv:hep-ph/9604311.

[58] J.C. Collins, "Scaling behavior of ϕ^4 theory and dimensional regularization," Phys. Rev. D **10**, 1213–1218 (1974).

[59] H. Georgi and S.L. Glashow, "Unity of all elementary-particle forces," Phys. Rev. Lett. **32**, 8 (1974).

[60] A. Linde, *Particle physics and inflationary cosmology* (arXiv:hep-th/0503203).

[61] E.W. Kolb and M.S. Turner, *The early universe* (Addison–Wesley Publishing Company 1993).

[62] S.E. Rugh and H. Zinkernagel, "The quantum vacuum and the cosmological constant problem," Stud. Hist. Philos. Mod. Phys. **33**, 663 (2002); arXiv:hep-th/0012253.

[63] V. Rubakov, *Classical theory of gauge fields* (Princeton University Press 1999).

[64] CMS Collaboration, http://cdsweb.cern.ch/record/1460438.

[65] ATLAS Collaboration, http://cdsweb.cern.ch/record/1460439.

[66] J. Dreitlein, "Broken symmetry and the cosmological constant," Phys. Rev. Lett. **33**, 1243–1244 (1974).

[67] M. Veltman, "Cosmology and the Higgs mass," Phys. Rev. Lett. **34**, 777 (1975).

[68] S. Weinberg, "The cosmological constant problem," Rev. Mod. Phys. **61**, 1 (1989).

[69] L. Susskind, "Dynamics of spontaneous symmetry breaking in the Weinberg-Salam theory," Phys. Rev. D **20**, 10, 2619–2625 (1979).

[70] G. 't Hooft, in *Recent developments in gauge theories* (Proceedings of the NATO Advanced Summer Institute, Cargese 1979).

[71] Y. Nambu and G. Jona-Lasinio, "Dynamical model of elementary particles based an an analogy with superconductivity," Phys. Rev. **122**, 1 (1961).

[72] E.V. Shuryak, *The QCD vacuum, hadrons and superdense matter* (Second Edition, World Scientific Publishing Co. Pte. Ltd. 2004).

[73] (a) D.J. Gross and F. Wilczek, "Ultraviolet behavior of non-Abelian gauge theories," Phys. Rev. Lett. **30**, 26, 1343–1346 (1973); (b) H.D. Politzer, "Reliable perturbative results for strong interactions?," Phys. Rev. Lett. **30**, 26, 1346–1349 (1973).

[74] M.A. Shifman, A.I. Vainshtein and V.I. Zakharov, "QCD and resonance physics. Theoretical foundations," Nucl. Phys. B **147**, 385–447 (1978); "QCD and resonance physics. Applications," Nucl. Phys. B **147**, 448–518 (1978).

[75] S.J. Brodsky and R. Shrock, "Condensates in quantum chromodynamics and the cosmological constant," Proc. Natl. Acad. Sci. **108**, 45–50 (2011); arXiv:hep-th/0905.1151.

[76] H. Reinhardt and H. Weigel, "The vacuum nature of the QCD condensates," Phys. Rev. D **85**, 074029 (2012); arXiv:hep-ph/1201.3262.

[77] V. Sahni and A. Starobinsky, "The case for a positive cosmological Λ-term," IJMPD **9**, 373–444 (2000); arXiv:astro-ph/9904398.

[78] A. Vilenkin, "Cosmological constant problem and their solutions," arXiv:hep-th/0106083.

[79] S. Nobbenhuis, "Categorizing different approaches to the cosmological constant problem," Found. Phys. **36**, 5, 613–680 (2006); arXiv:gr-qc/0411093.

[80] F.R. Klinkhamer and G.E. Volovik, "Dynamics of the quantum vacuum: Cosmology as relaxation to the equilibrium state," Journal of Physics: Conference Series **314**, 012004 (2011); arXiv:gr-qc/1102.3152.

[81] G.E. Volovik, "From analogue models to gravitating vacuum," arXiv:gr-qc/1111.1155.

[82] L.H. Ford, "Cosmological-constant damping by unstable scalar fields," Phys. Rev. D **35**, 2339–2344 (1987).

[83] R.W. Hellings *et. al.*, "Experimental test of the variability of G using Viking Lander Ranging date," Phys. Rev. Lett. **51**, 18, 1609–1612 (1983).

[84] J.G. Williams, X.X. Newhall and J.O. Dickey, "Relativity parameters determined from lunar laser ranging," Phys. Rev. D **53**, 12, 6730–6739 (1996).

[85] F.R. Klinkhamer and G.E. Volovik, "Dynamical vacuum variable and equilibrium approach in cosmology," Phys. Rev. D **78**, 063528 (2008); arXiv:gr-qc/0806.2805.

[86] F.R. Klinkhamer and G.E. Volovik, "$f(R)$ cosmology from q-theory," JETP Letters, **88**, 5, 289–294 (2008); arXiv:gr-qc/0807.3896.

[87] M.J. Duff and P. van Nieuwenhuizen, "Quantum inequivalence of different field representations," Phys. Lett. B **94**, 2, 179–182 (1980).

[88] A. Aurilia, H. Nicolai and P.K. Townsend, "Hidden constants: The θ parameter of QCD and the cosmological constant of $N = 8$ supergravity," Nucl. Phys. B **176**, 2, 509–522 (1980).

[89] R. Jackiw, "4-dimensional Einstein gravity extended by a 3-dimensional gravitational Chern-Simons term," International Journal of Theoretical Physics **45**, 8, 1431–1441 (2006).

[90] F.R. Klinkhamer and G.E. Volovik, "Gluonic vacuum, q-theory, and the cosmological constant," Phys. Rev. D **79**, 063527 (2009); arXiv:gr-qc/0811.4347.

[91] V.A. Rubakov and P.G. Tinyakov, "Ruling out a higher spin field solution to the cosmological constant problem," Phys. Rev. D **61**, 087503 (2000); arXiv:hep-ph/9906239.

[92] C.M. Will and K. Nordtvedt, Jr., "Conservation laws and preferred frames in relativistic gravity. I. Preferred-frame theories and an extended PPN formalism," The Astrophysical Journal **177**, 757–774 (1972); "Conservation laws and preferred frames in relativistic gravity. II. Experimental evidence to rule out preferred-frame theories of gravity," The Astrophysical Journal **177**, 775–792 (1972).

[93] R.W. Hellings and K. Nordtvedt, Jr., "Vector-metric theory of gravity," Phys. Rev. D **7**, 3593–3602 (1973).

[94] C.M. Will, *Theory and experiment in gravitational physics* (Cambridge University Press 1993).

[95] (a) T. Jacobson and D. Mattingly, "Gravity with a dynamical preferred frame," Phys. Rev. D **64**, 024028 (2001); arXiv:gr-qc/0007031. (b) T. Jacobson, "Einstein-aether gravity: a status report," arXiv:gr-qc/0801.1547.

[96] D. Colladay and V.A. Kostelecky, "Lorentz-violating extension of the standard model," Phys. Rev. D **58**, 116002 (1998); arXiv:hep-ph/9809521.

[97] J.B. Jimenez and A.L. Maroto, "Cosmological evolution in vector-tensor theories of gravity," Phys. Rev. D **80**, 063512 (2009); arXiv:astro-ph/0905.1245.

[98] V. Emelyanov and F.R. Klinkhamer, "Vector-field model with compensated cosmological constant and radiation-dominated FRW phase," IJMPD **21**, 3, 1250025 (2012); arXiv:gr-qc/1108.1995.

[99] V. Emelyanov and F.R. Klinkhamer, "De-Sitter-spacetime instability from a nonstandard vector field," arXiv:gr-qc/1204.5085.

[100] A. Erdelyi, *Higher transcendental functions* (McGRAW-HILL Book Company 1953).

[101] V. Emelyanov and F.R. Klinkhamer, "Reconsidering a higher-spin-field solution to the main cosmological constant," Phys. Rev. D **85**, 063522 (2012); arXiv:hep-th/1107.0961.

[102] F.R. Klinkhamer, "Inflation and the cosmological constant," Phys. Rev. D **85**, 023509 (2012); arXiv:gr-qc/1107.4063.

[103] V. Emelyanov and F.R. Klinkhamer, "Possible solution to the main cosmological constant problem," Phys. Rev. D **85**, 103508 (2012); arXiv:hep-th/1109.4915.

[104] J. Setiawan *et. al.*, "Planetary companions around the metal-poor star HIP 11952," A&A **540**, A141 (2012); arXiv:astro-ph/1102.0503.

[105] M. Ostrogradsky, "Memoire sur les equations differentielles relatives au probleme des isoperimetres," Mem. Acad. St. Petersburg, **4**, 385–517 (1850).

[106] R.P. Woodard, "Avoiding dark energy with $1/R$ modifications of gravity," Lect. Notes Phys. **720**, 403–433 (2007); arXiv:astro-ph/0601672.

[107] A.D. Dolgov and M. Kawasaki, "Realistic cosmological model with dynamical cancellation of vacuum energy," arXiv:astro-ph/0307442.

[108] A.D. Dolgov and M. Kawasaki, "Stability of a cosmological model with dynamical cancellation of vacuum energy," arXiv:astro-ph/0310822.

[109] S.W. Hawking and G.F.R. Ellis, *The large scale structure of space-time* (Cambridge University Press 1973).

[110] B. Carter, "Energy dominance and the Hawking-Ellis vacuum conservation theorem," arXiv:gr-qc/0205010.

[111] S.M. Carroll, M. Hoffman and M. Trodden, "Can the dark energy equation-of-state parameter w be less than -1?," Phys. Rev. D **68**, 023509 (2003); arXiv:astro-ph/0301273.

[112] M. Visser and C. Barcelo, "Energy conditions and their cosmological implications," arXiv:gr-qc/0001099.

MIX
Papier aus verantwortungsvollen Quellen
Paper from responsible sources
FSC® C105338

Printed by Books on Demand GmbH, Norderstedt / Germany